Research Reports in Physics

Research Reports in Physics

Nuclear Structure of the Zirconium Region
Editors: J. Eberth, R. A. Meyer, and K. Sistemich

Ecodynamics Contributions to Theoretical Ecology
Editors: W. Wolff, C.-J. Soeder, and F. R. Drepper

Nonlinear Waves 1 Dynamics and Evolution
Editors: A. V. Gaponov-Grekhov, M. I. Rabinovich, and J. Engelbrecht

Nonlinear Waves 2 Dynamics and Evolution
Editors: A. V. Gaponov-Grekhov, M. I. Rabinovich, and J. Engelbrecht

Nonlinear Waves 3 Physics and Astrophysics
Editors: A. V. Gaponov-Grekhov, M. I. Rabinovich, and J. Engelbrecht

Nuclear Astrophysics
Editors: M. Lozano, M. I. Gallardo, and J. M. Arias

Optimized LCAO Method and the Electronic Structure of Extended Systems
By H. Eschrig

Nonlinear Waves in Active Media
Editor: J. Engelbrecht

Problems of Modern Quantum Field Theory
Editors: A. A. Belavin, A. U. Klimyk, and A. B. Zamolodchikov

Fluctuational Superconductivity of Magnetic Systems
By M. A. Savchenko and A. V. Stefanovich

Nonlinear Evolution Equations and Dynamical Systems
Editors: S. Carillo and O. Ragnisco

Nonlinear Physics
Editors: Gu Chaohao, Li Yishen, and Tu Guizhang

Nonlinear Waves in Waveguides with Stratification
By S. B. Leble

Quark–Gluon Plasma
Editors: B. Sinha, S. Pal, and S. Raha

Symmetries and Singularity Structures Integrability and Chaos in Nonlinear Dynamical Systems
Editors: M. Lakshmanan and M. Daniel

Modeling Air-Lake Interaction Physical Background
Editor: S. S. Zilitinkevich

M. Lakshmanan M. Daniel (Eds.)

Symmetries and Singularity Structures

Integrability and Chaos in Nonlinear Dynamical Systems

Proceedings of the Workshop, Bharathidasan University, Tiruchirapalli, India, November 29–December 2, 1989

With 23 Figures

Springer-Verlag
Berlin Heidelberg New York London
Paris Tokyo Hong Kong Barcelona

Professor Dr. Muthuswamy Lakshmanan
Dr. Muthiah Daniel
Department of Physics, Bharathidasan University, Palkalaiperur Campus,
Tiruchirapalli – 620 024, Tamilnadu, India

ISBN-13:978-3-540-53092-3 e-ISBN-13:978-3-642-76046-4
DOI: 10.1007/978-3-642-76046-4

2157 / 3140-543210 – Printed on acid-free paper

Preface

Symmetries and singularity structures play important roles in the study of nonlinear dynamical systems. It was Sophus Lie who originally stressed the importance of symmetries and invariance in the study of nonlinear differential equations. However, the full potentialities of symmetries had been realized only after the advent of solitons in 1965. It is now a well-accepted fact that associated with the infinite number of integrals of motion of a given soliton system, an infinite number of generalized Lie Bäcklund symmetries exist. The associated bi-Hamiltonian structure, Kac-Moody, Virasoro algebras, and so on, have been increasingly attracting the attention of scientists working in this area. Similarly, in recent times the role of symmetries in analyzing both the classical and quantum integrable and nonintegrable finite dimensional systems has been remarkable.

On the other hand, the works of Fuchs, Kovalevskaya, Painlevé and coworkers on the singularity structures associated with the solutions of nonlinear differential equations have helped to classify first and second order nonlinear ordinary differential equations. The recent work of Ablowitz, Ramani and Segur, conjecturing a connection between soliton systems and Painlevé equations that are free from movable critical points, has motivated considerably the search for the connection between integrable dynamical systems with finite degrees of freedom and the Painlevé property. Weiss, Tabor and Carnevale have extended these ideas to partial differential equations. Very recently, it has been found that the singularity structure of solutions in the complex time plane contains very important information regarding nonintegrability aspects as well.

In order to take stock of the present understanding on the above aspects, a workshop on "Symmetries and Singularity Structure Aspects of Nonlinear Dynamical Systems" was held at Bharathidasan University, Tiruchirapalli during November 29 – December 2, 1989. This book, containing nineteen articles, is the proceedings of the workshop.

The book consists of three parts. Part I deals with the symmetry aspects of nonlinear dynamical systems. Generalized symmetries, Lie, Lie-Bäcklund and master symmetries, Virasoro, Kac–Moody algebras and constants of motion of soliton systems are described in this part. Part II discusses the singularity structure analysis of Hamiltonian and non-Hamiltonian systems. Part III deals with the symmetry and integrability of classical and quantum chaotic systems. Quantum groups, random matrix theory, integrable spin chains and solitons in plasma, ^{4}He films, Josephson junctions and also chemical compounds are the topics covered in this part.

We are grateful to Dr. S. Muthukumaran, Vice-Chancellor, and to Professor P.K. Ponnuswamy for their support and encouragement. The workshop was generously sponsored by the International Centre for Theoretical Physics (ICTP), Trieste, Italy, the Department of Science and Technology of the Government of India, the Indian National Science Academy, and Bharathidasan University. Our colleagues at the Department of Physics have been extremely helpful in this endeavor for which we are very thankful. Finally, the neat typing of the manuscript by Mr. S. Venugopal is happily acknowledged.

Tiruchirapalli, India
July 1990

M. Lakshmanan
M. Daniel

Contents

Part III Integrability and Chaos: Quantum and Classical

Part I

Symmetry Aspects

Symmetries, Singularities and Exact Solutions for Nonlinear Systems

P. Winternitz

Centre de Recherches Mathématiques, Université de Montréal,
CP 6128-A, Montréal, Québec, Canada, H3C 3J7

The method of symmetry reduction for nonlinear partial differential equations and its combination with singularity analysis is reviewed. Physical examples, such as the quintic multidimensional Schrödinger equation, are discussed.

1. Introduction

It has become abundantly clear over the last few years that nonlinear phenomena in physics have finally caught up with us. Indeed, virtually all the fundamental equations of physics are either nonlinear, or involve nonlinear boundary conditions. It suffices to mention the Einstein equations of gravitation theory, the Yang-Mills equations of elementary particle theory, the Navier-Stokes, or Euler equations of hydrodynamics, and any number of others.

Nonlinear equations are notoriously difficult to solve. Several different approaches have proven themselves to be useful; the most standard one is to impose appropriate boundary conditions and then to integrate numerically. The second is to simplify the equations, making certain physical assumptions, and to reduce them to linear equations, or less extremely, to "integrable" nonlinear equations. Prime examples of this approach are the derivation of such equations as the Korteweg-de Vries equation, the Boussinesq equation, the Kadomtsev-Petviashvili equation and many others from the Euler equations. The obtained equations can then be solved by essentially linear techniques, such as the inverse scattering transform [1].

The purpose of this contribution is to discuss a third approach to nonlinear differential equations. We have in mind a systematic application of Lie group theory to obtain classes of exact particular solutions of nonlinear partial differential equations. The use of group theory in this context is anything but new. However, an avalanche of recent books, articles and conference proceedings on group theory and differential equations testifies to the topicality of this subject [2-8]. This resurgence is due to many factors, not the least of which is the advent of computer algebra. This makes it possible to perform

Research Reports in Physics Symmetries and Singularity Structures
Editors: M. Lakshmanan · M. Daniel

calculations, traditionally considered to be very cumbersome, with the aid of computers, using languages such as MACSYMA, REDUCE, MATHEMATICA and many others. Another development in this area is the rapid progress made in the theory of infinite dimensional Lie algebras, in particular Kac-Moody algebras [9,10]. New tools have been made available for clarifying subalgebras of Lie algebras [11,12,7]. Finally, a combination of group theory with singularity analysis, in particular Painlevé analysis, turns out to be very fruitful [13-17].

We shall concentrate here on just one of the many aspects of applying group theory to differential equations, namely the method of symmetry reduction. This is used to reduce the number of independent variables in an equation, in the extreme case to reduce a partial differential equation (pde) to an ordinary differential equation (ode), or even to an algebraic one.

2. Outline of the Method of Symmetry Reduction

We shall restrict ourselves here to the use of local Lie point symmetries of differential systems. These are transformations of the form

$$x' = \Lambda_g(x,u), \qquad u' = \Omega_g(x,u) \tag{1}$$

where $x = (x_1,\dots,x_p)$, $u = (u_1,\dots,u_q)$ are the independent and dependent variables, respectively and g denotes the group parameters. If $u(x)$ is a solution of the considered differential system and the functions $\Lambda_g(x,u)$ and $\Omega_g(x,u)$ are defined, then $u'(x')$ is also a solution. Thus, the set of all solutions is invariant and individual solutions are transformed amongst each other.

We write the system as

$$\Delta^\mu(x,u,u_{x_2}, u_{x_i x_\epsilon},\dots) = 0, \quad \mu = 1,\dots,n \tag{2}$$

that is, we have in general n equations, q unknowns, p independent variables and any (finite) number of partial derivatives.

The method of symmetry reduction consists of the following steps.

A. Find the symmetry group G of the equations

This is best done by using infinitesimal techniques. We actually construct the Lie algebra L of the Lie group G, realized by vector fields of the form

$$\hat{v} = \sum_{i=1}^{p} \xi_i(x,u)\partial_{x_i} + \sum_{a=1}^{q} \varphi_a(x,u)\, \partial_u a. \tag{3}$$

If the system (2) is of order N (involves derivatives of the order N) then we construct the N-th prolongation of v, i.e. the differential operator

$$pr^N\hat{v} = \hat{v} + \sum_{a,i} \varphi_a^{x_i}(x,u,u_x)\partial_{u^a_{x_i}} + \sum_{a,i,k} \varphi^{x_i x_k}(x,u,u_x,u_{xx})\partial_{u^a_{x_i x_k}} + \ldots \quad (4)$$

The coefficients $\varphi_a^{x_i}, \ldots, \varphi^{x_{i_1} \cdots x_{i_N}}$ in the prolongation (4) are expressed in terms of the (unknown) functions ξ_i and φ_a in (3) and their derivatives upto order N. We skip the corresponding formulas involving total derivatives and refer e.g. to the books [2-7]. The coefficients $\xi_i(x,u)$ and $\varphi_a(x,u)$ are obtained by solving a system of linear partial differential equations: the so-called determining equations. These are in turn obtained from the requirement

$$pr^N\hat{v}.\Delta|_{\Delta=0} = 0 . \quad (5)$$

Various computer programs exist that construct the prolongation (4), read off the determining equations from (5) and then partly or completely solve them (for a MACSYMA program, see Ref. 18, for a REDUCE one, see Ref. 19).

The system of determining equations is linear, even if the studied equations (2) are nonlinear. It is in general overdetermined and may admit only a trivial solution ($\xi_i = 0$, $\varphi_a = 0$), meaning that G is just the identity transformation. In physically interesting cases the symmetry groups tend to be quite large. The integration of the determining equations introduces either a finite number of significant integration constants, or a finite number of arbitrary functions. Correspondingly, the symmetry algebra L may be finite or infinite dimensional.

Once the symmetry algebra L is found, the connected component of the symmetry group G is obtained by the standard procedure of integrating the now known vector fields (3). The corresponding one-parameter subgroups of G are obtained by solving the equations

$$\frac{dx_i'}{d\lambda} = \xi_i(x',u'), \quad \frac{du^{a'}}{d\lambda} = \varphi_a(x',u'), \quad x_i'|_{\lambda=0} = x_i, \quad u^{a'}|_{\lambda=0} = u^a. \quad (6)$$

B. Identify the symmetry algebra L

The symmetry algebra L is realized by vector fields (3). It is easy to choose some basis and to calculate the structure constants in this basis. To proceed further, we need to identify L as a specific abstract Lie algebra. Algorithms are available [20] (and have been realized) as computer packages [21,22] that decompose L into a direct sum $L = L_1 \oplus L_2 \oplus \ldots$, if L is decomposable.

For the indecomposable components they determine whether they are simple, solvable, or semidirect sums of the form $L = S \triangleright R$, where S is semisimple and R is the radical (maximal solvable ideal). The algorithms provide the Levi decomposition [23,24] whenever it is nontrivial. For every solvable component they find the nilradical (maximal nilpotent ideal).

In other words, the symmetry algebra L is transformed to a canonical basis in which all of its basic properties are manifest.

C. Classify the subalgebras of L into conjugacy classes

The classification is to be performed under the action of the symmetry group G and a representative of each class is to be chosen. This classification of subalgebras of L also provides a classification of closed connected subgroups of G. Each representative subgroup $H \subset G$ will provide a different reduction of the considered system (2) to a lower dimensional system, provided H satisfies certain conditions. The symmetry group G and hence its subgroups H act on the space $M \sim X \times U$ of independent and dependent variables. We need to find all subgroups H having generic orbits of codimension K+q in M, where $q = \dim U$, and k satisfies $0 \leq k \leq p = \dim X$, such that the projection of the orbits onto X has codimension k.

Methods have been developed for classifying subalgebras of finite dimensional Lie algebras into conjugacy classes [11,12]. They have been extended to the classification of finite dimensional subalgebras of certain types of infinite dimensional Lie algebras [25,26]. For a recent review see Ref. [7]. Let us just mention that if L is a simple Lie algebra, then the methods to be used are extensions of the results of Dynkin [27], Karpelevich [28] and Cornwell [29]. If L is a direct sum, the method is an adaptation of Goursat's classical method [30]. For semidirect, sums, $L = F \triangleright N$, see Ref. [11,12],7].

D. Find the invariants of each representative subgroup H

If H satisfies the conditions stated above, there will exist k+q functionally independent invariants $I_\mu(x,u)$ on $M \sim X \times U$, such that k of them can be chosen to be functions on X alone $\eta_1(x),\dots,\eta_k(x)$, whereas the remaining ones, $F_1(x,u),\dots,F_q(x,u)$ satisfy a nondegeneracy condition

$$\det\left(\frac{\partial F_\mu}{\partial u^k}\right) \neq 0 . \qquad (7)$$

Determining $\eta_i(x)$ and $F_j(x,u)$ involves the solution of a system of linear first order partial differential equations.

E. Perform the reduction

The structure of the invariants makes it possible to express the dependent variables in terms of the invariants F_i

$$u^a(x) = U^a(F_1,\ldots,F_q,x_1,\ldots,x_p) \ . \tag{8}$$

We then view the quantities F_j as functions of the remaining invariants x_i, i.e. put $F_j = F_j(\xi_1,\ldots,\xi_k)$, $1 \le j \le q$. Substituting (8) back into the original system (2) we obtain a system of equations for the functions F_j, involving only $\xi_1,\ldots,\xi_k$ as independent variables. Invariance of the original system, together with the completeness of the set of invariants assures that the original variables x_i drop out. Since we have $k < p$ we always have a dimensional reduction. For $k = 0$ we obtain an algebraic equation, for $k = 1$ an ode.

F. *Solve the reduced system*

The reduced system is usually also a nonlinear one and may be difficult to solve. In some cases, the reduced system is integrable by inverse scattering techniques, even if the original one is not. If the reduction goes all the way down to an ode, then the most fruitful approach is usually to perform a singularity analysis and to pick out, amongst the reduced odes those that have the Painlevé property (no movable singularities, other than poles) [13-17].

Odes that have the Painlevé property are in general easier to solve than those that do not. Moreover, equations with this property can be classified into conjugacy classes under Möbius transformations. Thus if an n-th order ode for y(t) has the Painlevé property, then so does the n-th order ode for z(x) where

$$z(x) = \frac{\alpha(t)y(t) + \beta(t)}{\gamma(t)y(t) + \delta(t)} \ , \quad x = \varphi(t) \ , \tag{9}$$

and α, β, γ, δ and φ are arbitrary sufficiently smooth functions. Such classifications have been performed for several cases, namely:

(i) $\dot{y} = f(y,t)$ (10)

where f is rational in y and analytic in t.

All such equations with the Painlevé property are transformable into the Riccati equation [31]

(ii) $A(y,t)\dot{y}^2 + B(y,t)\dot{y} + C(y,t) = 0$ (11)

where A, B and C are polynomials in y and analytic in t. Such equations are solvable in terms of solutions of the Riccati equation, elliptic functions, or elementary functions that are degenerate cases of elliptic functions [31]

(iii) $\ddot{y} = f(\dot{y},y,t)$ (12)

with f rational in $\dot{y}$ and y and analytic in t. These are the equations classified by Painlevé and Gambier [13,14,31] into 50 classes, 6 of which provide the now famous Painlevé transcendents. The others can be integrated in terms of known special functions (mainly elliptic functions).

$$\text{(iv)} \quad A(\dot{y},y,t)\ddot{y}^2 + B(\dot{y},y,t)\ddot{y} + C(\dot{y},y,t) = 0 \tag{13}$$

where A, B, and C are polynomials in $\dot{y}$ and y and analytic in t. Such equations have been classified and integrated by Bureau [32] and collaborators [33], also in terms of the Painlevé transcendents and other known functions.

$$\text{(v)} \quad \dddot{y} = F(\ddot{y},\dot{y},y,t) \tag{14}$$

where F is rational in $\ddot{y}$, $\dot{y}$ and y and analytical in t. A partial classification of such equations and their reduction to lower order ones is also due to Bureau [34].

3. Examples

(1) The Infinite Dimensional Symmetry Algebra of the Kadomtsev-Petviashvili Equation

The Kadomtsev-Petviashvili equation describes the propagation of weakly nonlinear waves in a dispersive two-dimensional medium. It can be written as

$$(u_t + \frac{3}{2}uu_x + \frac{1}{4}u_{xxx})_x + \frac{3}{4}\alpha u_{yy} = 0 \, , \quad \alpha = \pm 1 \tag{15}$$

where u is the wave amplitude. This is the prototype equation in 2+1 dimensions, integrable by the inverse spectral transform [1]. Applying the algorithm described above (and the MACSYMA program [18]) we find that the general element of the symmetry algebra can be written as:

$$V = T(f) + Y(g) + X(h) \tag{16}$$

where f(t), g(t) and h(t) are arbitrary C^∞ functions of time t and

$$\begin{aligned} T(f) &= f(t)\partial_t + [\frac{1}{3}x\dot{f} - \frac{2}{9}\alpha y^2\ddot{f}]\partial_x + \frac{2}{3}y\dot{f}\partial_y \\ &\quad - [\frac{4}{27}\alpha y^2 \dddot{f} - \frac{2}{9}\ddot{f} + \frac{2}{3}u\dot{f}]\partial_u \\ Y(g) &= g(t)\partial_y - \frac{2}{3}\alpha y\dot{g}\partial_x - \frac{4\alpha}{9}y\ddot{g}\partial_u \\ X(h) &= h(t)\partial_x + \frac{2}{3}\dot{h}(t)\partial_u \, . \end{aligned} \tag{17}$$

For the corresponding group transformations, the embedding of the Lie algebra (16) into a Kac-Moody-Virasoro algebra and for applications to symmetry reduction, we refer to the original articles [25]. Let us just mention that the obvious "physical" transformations are obtained by restricting the arbitrary functions in (17) to be linear polynomials. Thus, T(1), Y(1) and X(1) generate time and space translations, T(t) generates dilations, Y(t) a shear transformation, i.e.

$$x' = x - \frac{2}{3}\alpha\lambda y - \frac{1}{3}\alpha\lambda^2 t \ , \quad y' = y + \lambda t, \quad t' = t, \quad u' = u \ ,$$

and X(t) a Galilei transformation in the x direction.

The symmetry algebras of other integrable equations involving 3 independent variables, such as the Davey-Stewartson equation [35], the 3-wave resonant interaction equations [26] and many others, are also infinite dimensional and have a typical Kac-Moody-Virasoro structure [6].

(2) Symmetry Reduction for a Nonlinear Schrödinger Equation

Let us now consider a nonintegrable nonlinear evolution equation, namely the nonlinear Schrödinger equation

$$i\psi_t + \Delta\psi = |\psi|^n\psi \ , \ n \neq 0 \ , \ \psi = Ne^{i\alpha}, \quad 0 \leq N < \infty, \ 0 \leq \alpha \leq 2\pi \quad (18)$$

in three space dimensions (it is integrable only in one space dimension and for n = 2). Let us now run through the symmetry reduction procedure in order to obtain particular solutions, compatible with boundary conditions imposed on a cylinder.

i. The symmetry algebra is the extended Galilei similitude algebra with basis:

$$P_i = \partial_{x_i} \ , \quad P_0 = \partial_t \ , \quad J_i = -\varepsilon_{ik\ell} x_k \partial_{x_\ell} \ , \quad K_i = t\partial_{x_i} + \frac{1}{2} x_i \partial_\alpha , M=-\partial_\alpha$$

$$D = 2t\partial_t + x_1\partial_{x_1} + x_2\partial_{x_2} + x_3\partial_{x_3} - \frac{2}{n} N\partial_N \quad (19)$$

where P_μ, J_i , M and D generate space and time translations, rotations, proper Galilei transformations, constant changes of phase and dilations, respectively.

ii. The symmetry algebra is already obtained in a canonical basis, i.e. we have the Levi decomposition

$$L\{J_1,J_2,J_3\} \ltimes \{D,P_0,K_i,P_i,M\} \quad (20)$$

where the radical is the expression on the right. The nilradical is given by $\{P_0,K_i,P_i,M\}$.

iii. The subalgebras of L of all dimensions have been classified completely [36]. Here we shall only present those that provide a reduction to an ode and that are compatible with a cylindrical geometry. They are represented by

$$L_1^{ab} = \{J_3 + aM,\ P_3,\ P_0 + bM\} \quad a \geq 0,\ b \in \mathbb{R} \tag{21}$$

$$L_2^{ab} = \{J_3 + aM,\ P_3,\ D + bM\} \quad a \geq 0,\ b \geq 0$$

$$L_3^{ab} = \{J_3 + aM,\ K_3,\ D + bM\} \quad a \geq 0,\ b \geq 0$$

where a, b are arbitrary real parameters in the indicated ranges.

iv. The invariants of the corresponding subgroups are easy to find. As an example we take L_1^{ab}, write the three basis elements as differential operators X_k using (19) and solve the pdes

$$X_k \Phi(x_i, t, M, \alpha) = 0 \quad , \ k = 1,2,3 . \tag{22}$$

We introduce polar coordinates $x_1 = \rho \sin\vartheta$, $x_2 = \rho \cos\vartheta$ and write the invariants as:

$$\xi = \rho = (x_1^2 + x_2^2)^{\frac{1}{2}} \ , \ F_1 = \alpha - a\vartheta + bt,\ F_2 = N . \tag{23}$$

v. The reduction formulas are

$$\psi(x_i, t) = f(\rho)\exp[i(a\vartheta - bt)], \quad f(\rho) = N(\rho)\exp[i\chi(\rho)] \tag{24}$$

$$\ddot{f} + \frac{1}{\rho}\dot{f} - \frac{a^2}{\rho^2} f = -bf + f|f|^n . \tag{25}$$

Restricting to n = 4 (quintic nonlinear Schrödinger equation) and separating out the real and imaginary parts, we obtain

$$\chi = S_0 \int \frac{d\rho}{\rho N^2} + \chi_0 , \tag{26}$$

$$\ddot{N} - \frac{S_0^2}{\rho^2 N^3} + \frac{1}{\rho}\dot{N} - \frac{a^2}{\rho^2} N = -bN + N^5 , \tag{27}$$

where S_0 and χ_0 are integration constants.

vi. In order to solve the rather complicated nonlinear ode (27) we first perform a Painlevé test [16], using the MACSYMA program [17]. We find that as it stands solutions of (27) will have algebraic branch points of the type $(\rho - \rho_0)^{-\frac{1}{2}}$. To avoid this we introduce an auxiliary function $H(\rho) = (N(\rho))^2$, satisfying

$$\ddot{H} = \frac{\dot{H}^2}{2H} - \frac{1}{\rho}\dot{H} + \frac{2a^2}{\rho^2} H + \frac{2S_0^2}{\rho^2 H} - 2bH + 2H^3 . \tag{28}$$

This equation passes the Painlevé test if and only if we have

$$a^2 = \frac{1}{16} , \quad b = 0 . \tag{29}$$

Restricting ourselves to these values of a^2 and b, we can transform (28) to its standard form [31] by putting

$$H(\rho) = \frac{1}{\sqrt{\rho}} W(\eta) , \quad \eta = \frac{4}{3}\sqrt{\rho} . \tag{30}$$

We obtain

$$\ddot{W} = \frac{\dot{W}^2}{2W} + \frac{3}{2} W^3 + -\frac{3}{2}\frac{S_0^2}{W} \tag{31}$$

i.e. one of the reducible Painlevé equations (PXXX, to be specific [31]).

Its first integral is

$$\dot{W}^2 = W^4 + 4CW - 3S_0^2 . \tag{32}$$

For $(C,S_0) = (0,0)$ we obtain an elementary solution

$$\psi(\mathbf{r},t) = \frac{1}{2}\left(\frac{3}{\rho}\right)^{1/4} \left(\frac{1}{\sqrt{\rho} + C_0}\right)^{1/2} \exp[i(\tfrac{\vartheta}{4} + X_0)] . \tag{33}$$

For $(C,S_0) \neq (0,0)$ we obtain a variety of different periodic solutions, all expressible in terms of Jacobi elliptic functions [37,38].

The conditions (29) point to the importance of a systematic subgroup classification procedure. Had we simply imposed cylindrical invariance by choosing the subalgebra $L_1 = \{J_3,P_3,P_0\}$ we would have obtained a reduction to a non Painlevé type equation for which no explicit solutions are available.

4. Conclusions

The emphasis in this contribution was on the construction of explicit analytic solutions of nonlinear differential systems. Left aside was the question of their applications. Obvious questions to ask are whether the obtained solutions are stable, whether they are asymptotic solutions, serving at least as local attractors, etc. In any case, exact solutions, obtained for particular initial and boundary conditions, can serve to test numerical procedures, can form the basis for further perturbative calculations, and most important, can clarify the qualitative behavior of nonlinear systems.

Acknowledgements

The author's research was partially supported by research grants from NSERC of Canada and FCAR du Québec. He thanks Professor M. Lakshmanan of Bharathidasan University for the opportunity to present this material.

References

[1] M. J. Ablowitz and H. Segur, **Solitons and the Inverse Scattering Transform** (SIAM, Philadelphia, 1981).

[2] P. J. Olver, **Applications of Lie Groups to Differential Equations** (Springer-Verlag, New York, 1986).

[3] G. W. Bluman and S. Kumei, **Symmetries and Differential Eqqtions** (Springer-Verlag, New York, 1989).

[4] L. V. Ovsiannikov, **Group Analysis of Differential Equations** (Academic Press, New York, 1982).

[5] N. H. Ibragimov, **Transformation Groups Applied to Mathematical Physics** (Reidel, Boston, 1985).

[6] D. Levi and P. Winternitz, **Symmetries and Nonlinear Phenomena** (World Scientific, Singapore, 1988).

[7] P. Winternitz, in **Partially Integrable Nonlinear Evolution Euqations and Their Physical Applications** (Kluwer, Dordrecht, 1990).

[8] A. M. Vinogradov et al., **Special Issues on Symmetries of Partial Differential Equations,** Acta Applicandae Math. **15** and **16** (1989).

[9] V. G. Kac, **Infinite Dimensional Lie Algebras** (Birkhauser, Boston, 1983).

[10] P. Goddard and D. Olive, Int. J. Mod. Phys. **A1** (1986) 303.

[11] J. Patera, P. Winternitz and H. Zassenhaus, J. Math. Phys. **16** (1975) 1597; **16** (1975) 1615; **17** (1976) 717.

[12] J. Patera, R. T. Sharp, P. Winternitz and H. Zassenhaus, J. Math. Phys. **17** (1976) 977; **18** (1977) 2259.

[13] P. Painlevé, Acta Math. **25** (1902) 1.

[14] B. Gambier, Acta Math. **33** (1910) 1.

[15] F. J. Bureau, Ann. Mat. Pura Appl. **IV 64** (1964) 229.

[16] M. J. Ablowitz, A. Ramani and H. Segur, J. Math. Phys. **21** (1980) 715; **21** (1980) 1006.

[17] D. Rand and P. Winternitz, Comp. Phys. Commun. **42** (1986) 359.

[18] B. Champagne and P. Winternitz, Preprint CRM-1278 (1985).

[19] F. Schwarz, Computing **34** (1985) 91.

[20] D. Rand, P. Winternitz and H. Zassenhaus, Lin. Alg. Applic. **109** (1988) 197.

[21] D. Rand, Comp. Phys. Commun. **41** (1986) 105.

[22] D. Rand, P. Winternitz and H. Zassenhaus, Comp. Phys. Commun. **46** (1987) 297.

[23] H. Zassenhaus, **Lie Groups, Lie Algebras and Physical Applications** (Presses de I'Université de Montréal, Montréal, 1981).

[24] N. Jacobson, **Lie Algebras** (Dover, New York, 1979).

[25] D. David, N. Kamran, D. Levi and P. Winternitz, Phys. Rev. Lett. **55** (1985) 2111; J. Math. Phys. **27** (1986) 1225.

[26] L. Martina and P. Winternitz, Ann. Phys. **196** (1989) 231.

[27] E. B. Dynkin, Ann. Math. Soc. Transl. Ser. **26** (1957) 245.

[28] F. I. Karpelevich, Trudy Mosk. Mat. Obsach. **4** (1955) 3.

[29] J. F. Cornwell, **Group Theory in Physics**, Vols. 1,2 (Academic Press, London, 1984).

[30] E. Goursat, Ann. Sci. Ec. Normale Sup. (3) **6** (1889) 9.

[31] E. L. Ince, **Ordinary Differential Equations** (Dover, New York, 1956)

[32] F. J. Bureau, Ann. Mat. Pura Appl. IV **41** (1972) 164.

[33] F. J. Bureau, A. Garcet and J. Goffar, Ann. Mat. Pura Appl. IV **92** (1972) 177.

[34] F. J. Bureau, Ann. Mat. Pura Appl. IV **66** (1964) 1.

[35] B. Champagne and P. Winternitz, J. Math. Phys. **29** (1988) 1.

[36] L. Gagnon and P. Winternitz, J. Phys. **A21** (1988) 1493; **A22** (1989) 469.

[37] L. Gagnon and P. Winternitz, Phys. Rev. **A39** (1989) 296.

[38] L. Gagnon and P. Winternitz, Phys. Lett. **A134** (1989) 276.

Application of Isovector Approach for the Solutions of Differential Equations of Physical Systems

O.P. Bhutani[1] *and K. Vijaya Kumar*[2]

[1]Department of Mathematics, Indian Institute of Technology, New Delhi 110016, India
[2]Department of Mathematics, Punjab University, Chandigarh 160014, India

A generalized geometric procedure for investigating the symmetries of differential equations namely 'isovector method' has been outlined and its application to obtain the solutions of the generalized diffusion equation and the semilinear hyperbolic equation with special reference to Liouville equation is indicated.

1. Introduction

Over the years quite a number of techniques have been employed to investigate Lie point transformation or symmetries admitted by the differential equations. Particular mention may be made of the works of Bluman and Cole [1], Ovsiannikov [2], Edelen [3] and Olver [4]. Even though these methods are algorithmic, the need for more generalized Lie point transformations, which include classical and non-classical, local and non-local is obvious. One such method termed as 'Isovector Approach' is due to Harrison and Estabrook [5] and seems to have hardly been exploited. The said approach is an alternative and systematic method to obtain the generators of the symmetries of differential equations. Based on the fundamental paper by Harrison and Estabrook [5], Edelen [6] presented a version that eventually became responsible for further development and its carry over to a number of problems in continuum mechanics. It is based on the vector field of the dependent and independent variables over which the Lie derivative of the ideals of the Cartan's exterior differential forms of the differential equations remain invariant. These introduced fields are nothing but the isovector fields which generate one parameter groups of invariant transformations with suitable initial conditions. These one parameter groups of transformations are known as 'Isogroups' of the differential equations. Further, the isogroups map every known solution to one parameter family of solutions of Lie groups. In this aspect, isovector method is equivalent to infinitesimal transformation method due to Bluman and Cole [7]. Also, it is seen that isovector approach generates mappings from nonlinear differential equations to linear differential equations (Edelen [3]). In

Research Reports in Physics Symmetries and Singularity Structures
Editors: M. Lakshmanan · M. Daniel

certain situations, isovector method may be employed to generate conservation laws satisfied by the governing equations. Further, the isogroup obtained, via isovector method can be extended to non-local symmetries by prolonging the ideal of exterior differential forms of the differential equations (Papachristou [8]).

2. Isovector Method for Partial Differential Equations

Before we embark upon the entitled approach for partial differential equations (pdes), let us consider the following problem:

Given a pde or a set of pdes, does there exist an equivalent set of differential forms in involution with respect to the independent variables, that is, a set whose integral manifolds are solutions of the pdes? The answer to this is available in the works of Cartan [9], who has provided the criterion for the equivalence of a given set of pdes (in N-dependent variables and n-independent variables) with a closed set of differential forms in a differential manifold M. The said set of forms is the basis of a differential ideal of the algebra of forms on M. Such an integral manifold represents geometrically a solution of the original set of pdes.

In the following we discuss two specific applications of the above method, namely, (i) the generalized diffusion equation , and (ii) semilinear hyperbolic equation.

3. Generalised Diffusion Equation

3.1 The System

If T(x,t) denotes the temperature at a point, 'a' be an arbitrary constant, D_1, D_2 and D_3 are arbitrary functions of temperature T and b(x,t) is another arbitrary function of x and t, then the generalised diffusion equation under consideration can be expressed as

$$T_t = (D_1(T)T_x)_x + a(D_2(T))_x + b(x,t)D_3(T) . \tag{3.1}$$

Equation (3.1) has numerous applications in physical and engineering sciences (see Bhutani et al. [10]).

Following the procedure described in [10], we herein introduce the variables u and y as defined below:

$$u = T_t , \tag{3.2}$$

$$y = T_x . \tag{3.3}$$

Rewriting (3.1) in the language of exterior differential forms, we get

$$\alpha = dT - udt - ydx , \tag{3.4}$$

$$\beta = (u - y^2 D_1' - ayD_2' - bD_3)dx \wedge dt - D_1 dy \wedge dt, \quad (3.5)$$

$$d\alpha = -du \wedge dt - dy \wedge dx , \quad (3.6)$$

where α, β and $d\alpha$ are 1,2,2 forms, respectively, and $\wedge$ denotes exterior product of differential forms and $D_i' = dD_i/dT$, $i = 1,2,3$. It can be seen easily that the set of form (α, β, $d\alpha$) form a closed ideal.

Moreover, if $\alpha = 0, \beta = 0$ we arrive at (3.1)-(3.2). While for the case where only $d\alpha = 0$, it results in the integrability condition $T_{xt} = T_{tx}$.

3.2 Isovector Field

Define a vector field V over the spaces of 5-dimension, say E_5, whose coordinates are (x,t,T,u,y) and is of the form

$$V = V^t \partial_t + V^x \partial_x + V^u \partial_u + V^y \partial_y + V^T \partial_T , \quad (3.7)$$

where V^t, V^x, V^u, V^y, V^T are components of V in t,x,u,y,T directions respectively, and $\partial_t = \partial/\partial t$, $\partial_x = \partial/\partial x$ and so on. Vector field V defined by (3.7) is said to be an isovector field if

$$L_V I(\alpha, \beta, d\alpha) \subseteq I(\alpha, \beta, d\alpha), \quad (3.8)$$

where L_V is the Lie derivative over the vector field V.

3.3 Transport Property of Forms α, β

As mentioned in Sect. 2, the transport property of exterior differential form α and β can be expressed as

$$L_V(\alpha) = \delta\alpha , \quad (3.9)$$

$$L_V(\beta) = \xi\beta + W \wedge \alpha - \sigma d\alpha , \quad (3.10)$$

where δ, ξ and σ are arbitrary function of (x,t,T,u,y) and W is an arbitrary 1-form. Further

$$L_V(\alpha) = d(V \rfloor \alpha) + V \rfloor d\alpha , \quad (3.11)$$

$$L_V(\beta) = d(V \rfloor \beta) + V \rfloor d\beta , \quad (3.12)$$

where d denotes exterior differentiation and $\rfloor$ denotes inner multiplication of forms. Also, $d\beta$ occurring in (3.12) is a 3-form and is given as under:

$$d\beta = du\wedge dx\wedge dt - (2yD_1' + aD_2')dy\wedge dx\wedge dt - (y^2D_1'' + ayD_2'' + bD_3')dT\wedge dx\wedge dt - D_1' \, dT\wedge dy\wedge dt . \tag{3.13}$$

3.4 Calculation of Isovectors $(V^t, V^x, V^T, V^u, V^y)$

Assuming

$$V \rfloor \alpha = G , \tag{3.14}$$

and

$$W = Adt + Bdx + Cdu + Ddy + EdT , \tag{3.15}$$

where A, B, C, D and E are arbitrary functions of (x,t,T,u,y). On combining (3.4), (3.6), (3.11), (3.14) and (3.9) and performing expansion under exterior differentiation and inner-multiplication operations and collecting the coefficients of similar 1-forms and equating them to zero and eliminating from the resulting equations, we get a system of quasilinear partial differential equations involving V^t, V^x, V^T, V^u, V^y and are given as under:

$$V^t = -G_u , \quad V^x = -G_y , \quad V^T = G - uG_u - yG_y , \quad V^u = G_t + uG_T, \quad V^y = G_x + yG_T . \tag{3.16}$$

On making similar substitutions of (3.5), (3.6), (3.12), (3.13) and (3.15) in (3.10) and performing expansions as stated above and collecting the coefficients of similar 2-forms and equating them to zero, we get a system of equations which on elimination of the arbitrary functions ξ, σ, A, B, C, D and E occurring in them and on combining the resulting equations with systems of (3.16) yield the following set of equations (for details see [10]):

$$G_{uu} = 0 , \tag{3.17}$$

$$(u - y^2D_1' - ayD_2' - bD_3)G_{uy} + D_1(G_{ux} + yG_{uT}) = 0 , \tag{3.18}$$

$$\begin{aligned}(uG_T + G_t) &- (2yD_1' + aD_2')(G_x + yG_T) - (y^2D_1'' - ayD_2'' - bD_3') \\ &(G - uG_u - yG_y) - (u - y^2D_1' - ayD_2' - bD_3)G_{xy} - D_1(G_{xx} + yG_{xT}) \\ &- y(u - y^2D_1' - ayD_2' - bD_3)G_{yT} - yD_1(G_{xT} + yG_{TT}) \\ &- \frac{1}{D_1}(u - y^2D_1' - ayD_2' - bD_3)^2G_{yy} - \frac{D_1'}{D_1}(u - y^2D_1' - ayD_2' - bD_3) \\ &(G - uG_u - yG_y) - (u - y^2D_1' - ayD_2' - bD_3)(G_{xy} + G_T + yG_{yT}) = 0.\end{aligned} \tag{3.19}$$

After somewhat detailed algebraic manipulations, we can write down the solution to the above system in G and utilize it in the system (3.16) to obtain the following set of isovectors:

$$V^t = -(p - 1)a_1 t + a_2 , \quad V^x = -\frac{1}{2}(p - m - 1)a_1 x + a_3 ,$$

$$V^T = a_1 T , \quad V^y = a\,\frac{1}{2}(p - m + 1)y, \quad V^u = a_1 pu, \quad 2n = m+p+1 . \tag{3.20}$$

In writing (3.20) we have assumed $D_1(T) = T^m$, $D_2(T) = T^n$ and $D_3(T)= T^p$. Further, a_1, a_2 and a_3 occurring in (3.20) are arbitrary constants.

From the set of isovectors listed in (3.20) we arrive at the following invariance transformation under the choice $T'(0) = T$, $x'(0) = x$ and $t'(0) = t$:

$$\Theta = (x + \nu_3)/(t + \nu_2)^{(p-m-1)/2(p-1)} ,$$

$$T = (t + \nu_2)^{-1/(p-1)} \rho(\Theta) . \tag{3.21}$$

Equation (3.21) reduces (3.1) to the following ordinary differential equation:

$$\rho^m \rho'' + m\rho^{m-1}\rho'^2 + \frac{p-m-1}{2(p-1)}\Theta\rho' + an\rho^{n-1}\rho' + \frac{1}{p-1}\rho + b(\Theta)\rho^p = 0. \tag{3.22}$$

Equation (3.22) is not solvable unless the values of m,n,p, a and $b(\Theta)$ are specified. Further, using the procedure due to Bluman and Cole, we can write down exact solution to (3.22) which does not seem to have been noticed to date.

3.5 New Exact Solution

When $b(\Theta) = 0$, (3.22) is reduced to

$$\rho^m \rho'' + m\rho^{m-1}\rho'^2 + an\rho^{n-1}\rho' + \frac{p-m-1}{2(p-1)}\Theta\rho' + \frac{1}{p-1}\rho = 0. \tag{3.23}$$

Under the condition $m = -1$ with $p = m+3$ and $n = m+2$, a solution to (3.1) can be expressed as

$$T = \frac{2(t + \nu_2)}{2a(x + \nu_3)(t + \nu_2) + (x + \nu_3)^2 + 2c_1(t + \nu_2)^2} , \tag{3.24}$$

where c_1 is a constant of integration.

When $a = 0$, $b(\Theta) = 0$, (3.22) is reduced to the equation corresponding to the heat diffusion equation:

$$\rho^m \rho'' + m\rho^{n-1}\rho'^2 + \frac{p-m-1}{2(p-1)}\Theta\rho' + \frac{\rho}{p-1} = 0 . \tag{3.25}$$

Integrating (3.25) under the condition $(p-m-1)/2 = 1$, we get a solution to the heat diffusion equation as

$$T = \left[c_2(t + \nu_2)^{-m/(m+2)} - \frac{m}{m+2}\,\frac{(x + \nu_3)^2}{t + \nu_2}\right]^{1/m}, \tag{3.26}$$

where c_2 is a constant of integration.

4. Semilinear Hyperbolic Equation

We consider the semilinear hyperbolic equation of the form

$$T_{xt} = f(T)\ , \tag{4.1}$$

where $f(T)$ is an arbitrary function of T. The importance of this equation for different forms of $f(T)$ can be seen through its study by Barone et al. [11], Shiff [12], Calogero [13], Eisenhart [14] and Shadwick [15]. Herein we have considered two very important cases corresponding to $f(T) = (a + bT)^k$, $k \neq 0,1$ and $f(T) = e^{nT}$. In the last case, (4.2) assumes the form of Liouville equation. Following the procedure given in Sect. 3, we give below the isovectors, the similarity transformation and the exact solutions.

Case 1 $f(T) \doteq (a + bT)^k, \quad k \neq 0,1.$

For the case under consideration, after quite a detailed analysis, we have arrived at the following isovectors:

$$\begin{aligned}
V^x &= -(a_1x + a_2)\ ,\\
V^t &= -[a_1 - a_0(k-1)]t + a_3\ ,\\
V^T &= \frac{a_0}{b}(a + bT)\ ,\\
V^y &= (a_1 + a_0)y\ ,\\
V^u &= (a_0k - a_1)u\ ,
\end{aligned} \tag{4.2}$$

where a_0, a_1, a_2, a_3 are arbitrary constants. From (4.2) we arrive at the invariant transformation

$$\begin{aligned}
\theta &= \frac{x + \nu_3}{(t + \nu_4)^{-1/k}}\ ,\\
T &= -2/b + (t + \nu_4)^{-1/k}\ \rho(\theta)\ ,
\end{aligned} \tag{4.3}$$

where

$$\nu_3 = a_2/a_0\ , \quad \nu_4 = a_3\nu a_0k$$

and

$$\rho(\theta) = [b(1-k)^2]^{1/1-k}\theta^{1/1-k} .$$

When $k = 1$ in (4.2), we arrive at the isovectors corresponding to linear form of (4.1) which, eventually, helps us to write solution of (4.1) for $k = 1$, in term of Clifford-Bessel function [16].

Case 2 $f(T) = e^{nT}, \quad n \neq 0$

For this case, (4.1) reduces to

$$T_{xt} = e^{nT} , \quad n \neq 0 \tag{4.4}$$

which is termed as Liouville equation.

Following the procedure outlined in earlier sections we arrive at the following isovectors for the Liouville equation:

$$V^x = -(\tfrac{n}{2} a_1 x^2 + a_2 x + a_3) ,$$

$$V^t = -(\tfrac{n}{2} b_1 t^2 + b_2 t + b_3) ,$$

$$V^T = a_1 x + b_1 t + (a_2 + b_2)/n ,$$

$$V^y = a_1(1 + ny) + a_2 y ,$$

$$V^u = b_1(1 + nu) + b_2 u . \tag{4.5}$$

In (4.5), a_1, a_2, a_3, b_1, b_2 and b_3 are arbitrary constants. Corresponding invariant transformations are given as, for the choice $a_1=0=b_1$

$$\theta = \frac{x + \nu_1}{(t + \nu_2)^r} , \quad T = \frac{1}{n} [\log(t + \nu_2)^{-(1+r)} + \rho(\theta)] , \tag{4.6}$$

where

$$\nu_1 = (a_3/a_2) , \quad \nu_2 = (b_1/b_2) , \quad r = (a_2/b_2) \tag{4.7}$$

and

$$\rho(\theta) = \log[\frac{r}{2n\theta} \operatorname{sech}^2 \log(1/c\sqrt{\theta})] .$$

Thus, the solution to (4.4) can be expressed as

$$T = \frac{1}{n}\left[\log\left(r/2n \frac{1}{(x + \nu_1)(t + \nu_2)} \operatorname{sech}^2\left(\log \frac{(t+ \nu_2)^{r/2}}{c(x + \nu_1)^{\frac{1}{2}}}\right)\right)\right] , \tag{4.8}$$

where c is a constant of integration. Equation (4.8) represents soliton type of behaviour. For details of the derivation of solution one can refer to Bhutani et al. [17].

5. Conclusion

We have, in this review work, given the relevant analytical details for carrying over the isovector approach to the problems of physical interest. This has been possible through modified version of the work of Harrison and Estabrook [5] by Edelen [6]. The technique has been successfully carried over to the problems of generalised diffusion equation and semilinear hyperbolic equation and has resulted in a few new exact solutions with special reference to Liouville equation.

References

[1] G. W. Bluman and J. D. Cole, **Similarity Methods for Differential Equations** (Springer-Verlag, 1974).
[2] L. V. Ovsiannikov, **Group Analysis of Differential Equations** (Academic Press, 1982).
[3] D. G. B. Edelen, **Applied Exterior Calculus** (John Wiley, 1985).
[4] P. J. Olver, **Applications of Lie Groups to Differential Equations,** G.T.M. 107 (Springer-Verlag, 1986).
[5] B. K. Harrison and F. B. Estabrook, J. Math. Phys. **12**(1971)653.
[6] D. G. B. Edelen, Int. J. Engg. Sci. **20** (1982) 803.
[7] G. W. Bluman and J. D. Cole, J. Math. Mech. **18** (1969) 1025.
[8] C. J. Papachristou and B. K. Harrison, J. Math. Phys. **26** (1987) 1261.
[9] E. Cartan, **Lecons Surles Invariant Integraux** (Hermann, Paris, 1922).
[10] O. P. Bhutani and K. Vijayakumar, Int. J. Engg. Sci. **4** (1990) 326.
[11] A. Barone, F. Esposito, C. J. Magee and A. C. Scott, Rivista Nuovo Cimento **1** (1971) 227.
[12] L. I. Shiff, Phys. Rev. **84** (1951) 1.
[13] F. Calogero, Stud. Appl. Math. **70** (1984) 189.
[14] L. P. Eisenhart, **A Treatise on the Differential Geometry of Curves and Surfaces** (Boston, Mass. Ginn, 1909).
[15] W. F. Shadwick, J. Math. Phys. **19** (1978) 2312.
[16] G. M. Murphy, **Ordinary Differential Equation and Their Solutions** (Van-Nonstrand, 1960).
[17] O. P. Bhutani and K. Vijayakumar (under communication).

Master Symmetries of Certain Nonlinear Partial Differential Equations

R. Sarathy[1] *and K.M. Tamizhmani*[2]

[1]Department of Applied Mathematics, Indian Institute of Science, Bangalore 560012, India
[2]Department of Mathematics, Pondicherry University, Pondicherry 605104, India

In this paper, we discuss the existence of master symmetries for a class of nonlinear partial differential equations. Furthermore, we derive the associated time dependent symmetries. Finally, we analyse the Lie algebraic structure of classical, master and time-dependent symmetries.

1. Introduction

Symmetries play a very important role in deciding the integrability of nonlinear partial differential equations [1-5]. During the last two decades or so, Lie's one-parameter continuous point transformation has been effectively used to find a class of particular solutions of nonlinear pdes [1]. Further, this method has been used to reduce nonlinear pdes to nonlinear odes of Painlevé type, if the system is an integrable one.

Further invariance of nonlinear pdes can be obtained from the theory of Lie-Bäcklund (LB) symmetries [1-5]. For the past few years many authors have applied this technique to establish the integrability by deriving infinitely many commuting LB symmetries [1-11]. Recently, Fuchssteiner introduced [2] a new class of symmetries called master symmetries which involves space variables explicitly. As a further consequence of these symmetries, one can easily derive the associated class of time-dependent symmetries [5-10]. In this talk, we discuss in detail the various aspects of the symmetry analysis and show that this approach is an algorithmic and elegant way to prove the integrability of the system. We also point out the connection between conserved covariants and these symmetries. We finally demonstrate the method with certain important examples.

2. Lie-Bäcklund Symmetries

We consider an evolution equation in its abstract form

$$u_t = K(u) , \tag{2.1}$$

Research Reports in Physics Symmetries and Singularity Structures
Editors: M. Lakshmanan · M. Daniel

defined on a jet-space J^∞ of an infinite dimensional C^∞ manifold M; K(u) is a suitable C^∞ function defined on the real line with the boundary conditions that u and its spatial derivatives vanish rapidly at $x = \pm\infty$. Let M* be the dual of M.

Definition 1

Fréchet derivative. We deal with differentiable functions K(u) such that $K(u)\colon E_1 \to E_2$ between suitable vector spaces E_1 and E_2. By differentiable we always mean a notion of differentiability such that the derivative K'(u) is a linear map between E_1 and E_2 and such that the chain rule holds. If $v \in E_1$, the derivative of K(u) in the direction v is denoted by K'(u)[v] and is calculated by

$$K'(u)[v] = \left.\frac{\partial K(u + \varepsilon v)}{\partial \varepsilon}\right|_{\varepsilon=0} . \tag{2.2}$$

Definition 2

Symmetry. A function σ is a symmetry of Eq. (2.1) if it leaves (2.1) invariant within order ε, that is the equation $u_t + \varepsilon\sigma_t = K(u+\varepsilon\sigma)$ must be correct to order ε.

Using

$$\sigma_t = \sigma'[u_t] , \quad K(u + \varepsilon\sigma) = K(u) + \varepsilon K'[\sigma] + O(\varepsilon^2) ,$$

it follows that a symmetry σ must solve

$$[\sigma,K] = \sigma'[K] - K'[\sigma] = 0 . \tag{2.3}$$

If σ depends on t explicitly, then

$$\frac{\partial\sigma}{\partial t} + \sigma'[K] - K'[\sigma] = 0 . \tag{2.4}$$

Definition 3

Recursion operator. An operator valued function $\varphi\colon M \to M$ is called a recursion operator for (2.1) if

$$\varphi'[K] - K'\varphi + \varphi K' = 0.$$

Also the recursion relation $K_{m+n} = \varphi^n K_m$ holds. It is clear from this relation that a recursion operator maps symmetries onto symmetries. Such an operator φ is called hereditary if $\varphi'[\varphi x_1]x_2 - \varphi\varphi'[x_1]x_2$ is symmetric in x_1 and x_2 for all $x_1, x_2 \in M$.

Definition 4

The Fréchet derivative of an operator-valued function $\varphi(u)$ is defined in the following way:

$$\varphi'(u)[v]w = \frac{\partial\varphi}{\partial\varepsilon}(u+\varepsilon v)w\Big|_{\varepsilon=0},$$

where v and w are arbitrary functions. By using the standard LB analysis we can find the recursion operators for npdes if they exit.

In the next section, we discuss the existence of such operators for certain npdes. In the following, we use the notations $u_i = \partial^i u/\partial x^i$, $i = 1,2,\ldots$ and $D = D_x$, a total differential operator and $D^{-1}(.) = \int_{-\infty}^{x}(.)dx'$.

3. Master Symmetries and Their Lie Algebraic Structure

Let **M** be an infinite dimensional C^∞ manifold and let u be a variable running through **M**. Let L be a Lie algebra of C^∞ vectorfields and L be a suitable sub-algebra of L*. If A and B are C^∞-vector fields, then the Lie bracket [A,B] is given by

$$[A(u),B(u)] = \frac{\partial}{\partial\varepsilon}[A\{u+\varepsilon B(u)\} - B\{u+\varepsilon A(u)\}]\Big|_{\varepsilon=0} \tag{3.1}$$

$$= A'[B] - B'[A].$$

Consider some evolution equation

$$u_t = K(u), \quad u \in M \tag{3.2}$$

with $K \in L$.

Denote the adjoint map given for $B \in L^*$ by adB, that is,

$$\mathrm{ad}B[A] = [B,A].$$

For K, its commutant is,

$$K^\perp = \{A \in L / [A,K] = 0\}.$$

Definition 5

$T \in L^*$ is a master symmetry of degree 1 for Eq. (3.2) if

$$[T,A] \in K^\perp, \quad \text{for all } A \in K^\perp.$$

That is, adT maps K into itself. This definition implies

$$[[T,A],K] = 0, \quad \text{for all } A \in K^\perp$$

and $[T,A] \in L$, for all $A \in K^\perp$.

Definition 6

T is a master symmetry of degree n for Eq. (3.2) iff, for arbitrary $A_1,\ldots,A_n \in K^{\perp}$, we have

$$\mathrm{ad}A_1\mathrm{ad}A_2\ldots\mathrm{ad}A_n[T] \in K^{\perp}.$$

That is, if $A \in L$, then

$$[A, \mathrm{ad}A_1\mathrm{ad}A_2\ldots\mathrm{ad}A_nT] = 0 .$$

Definition 7

Let φ be the hereditary operator associated with the evolution equation (3.2). Then the function T_0 is a scaling symmetry of both K and φ if,

$$[K,T_0] = \alpha K,$$

$$\varphi'[T_0] + \varphi[T_0'] - T_0'[\varphi] = \beta\varphi ,$$

where $\alpha,\ \beta \in R$.

Theorem 3.1

Let K be a vector field and T_0 another vector field associated with a scaling symmetry of both K and of the hereditary operator φ. That is,

$$[K,T_0] = \alpha K , \tag{3.3}$$

$$\varphi'[T_0] + \varphi[T_0'] - T_0'[\varphi] = \beta\varphi, \tag{3.4}$$

where $\alpha,\ \beta \in R$.
Let

$$K_{m+n} = \varphi^n K_m , \quad T_n = \varphi^n T_0 , \quad n \in Z. \tag{3.5}$$

Then

$$[K_n, K_m] = 0 , \tag{3.6}$$

$$[K_m, T_n] = (\alpha + m\beta)K_{n+m} , \tag{3.7}$$

$$[T_m, T_n] = \beta(m-n)T_{n+m} , \tag{3.8}$$

and

$$\Sigma = (\alpha + n\beta)tK_{n+1} + \varphi T_0 \tag{3.9}$$

is a symmetry of $u_t = K_n(u)$.
For details of proof see ref.[5].

Note: φT_0 is a master symmetry of degree 1 for $u_t = K(u)$ and $\varphi^n T_0$ is a master symmetry of degree n for $u_t = K(u)$.
Using the above method in the following, we construct master and time-dependent symmetries for certain npdes and also we give their Lie bracket relations.

4. Examples

4.1 New Nonlinear Schrödinger Equation

$$K(u) = u_t = i(u_2 + 2uu_1 - 2u_1u^* - 2uu_1^*) \quad . \qquad (3.10)$$

Hereditary operator: $\varphi(u) = i[D + u_1D^{-1}\{(.)-(.)^*\} + u\{(.)-(.)^*\} - u\{u(.) + u^*(.)\}]$. These equations admit the scaling $u \to \alpha u$ and $x \to \alpha^{-1}x$. Then $T_0 = u + xu_1$. Also

$$[K,T_0] = 2K$$

$$\varphi'[T_0] + \varphi[T_0'] - T_0'[\varphi] = \varphi.$$

By theorem 3.1

$$[K_n, K_m] = 0$$

$$[K_m, T_n] = (2+m)K_{n+m}$$

$$[T_m,T_n] = (m-n)T_{n+m}$$

and

$$\Sigma = (2 + n)tK_{n+1} + \varphi T_0$$

is a symmetry of (3.10). Then $T = xK - i(3uu^* + 2uu_1 + u^2)$ is a master symmetry of degree 1 of Eq. (3.10).

4.2 A New Derivative Nonlinear Schrödinger Equation

$$K(u) = u_t = iu_2 - uu^*u_1 \quad . \qquad (3.11)$$

Hereditary operator $\varphi(u) = iD - \frac{1}{2}uu^* - \frac{1}{2}uD^{-1}\{u_1(.)^* - u_1^*(.)\} - \frac{1}{2}uD^{-1}\{u(.)^* + u^*(.)\}$. Here, $K(u)$ and $\varphi(u)$ admit the scaling $u \to \alpha u$, $x \to \alpha^{-2}x$. Then $T_0 = u + 2xu_1$. Also,

$$[K,T_0] = 4K$$

$$\varphi'[T_0] + \varphi[T_0'] - T_0'[\varphi] = 2\varphi \ .$$

By theorem 3.1,

$$[K_n, K_m] = 0$$

$$[K_m, T_n] = (4 + 2m)K_{n+m}$$

$$[T_m, T_n] = 2(m-n)T_{n+m}$$

and

$$\Sigma = (4 + 2n)tK_{n+1} + \varphi T_0$$

is a symmetry of (3.11). Then $T = 2xK - 3iu_1 + u^2u^*$ is a master symmetry of degree 1 of Eq. (3.11).

4.3 Modified Complex Burgers' Equation

$$K = u_t = iu_2 + u_1^2 \qquad (3.12)$$

$$\varphi = iD + u_1$$

$$T_0 = xu_1.$$

Also

$$[K, T_0] = 2K$$

$$\varphi'[T_0] + \varphi[T_0'] - T_0'[\varphi] = \varphi .$$

By theorem 3.1

$$[K_n, K_m] = 0$$

$$[K_m, T_n] = (2 + m)K_{n+m}$$

$$[T_m, T_n] = (m-n)T_{n+m}$$

and

$$\Sigma = (2 + n)tK_{n+1} + \varphi T_0$$

is a symmetry of (3.12). Then $T = xK - iu_1$ is a master symmetry of degree 1 of Eq. (3.12).

We wish to remark that the conserved quantities f_n's can also be obtained recursively using the master symmetries T_m:

$$\langle \text{grad } f_n, T_m \rangle = (m + n)f_{n+m} .$$

Acknowledgment

The authors express their sincere gratitude to Professor M. Lakshmanan for giving us the constant encouragement and valuable suggestions.

References

[1] M. Lakshmanan and P. Kaliappan, J. Math. Phys. **24** (1983) 795.
P. J. Olver, **Applications of the Lie Groups to Differential Equations** (Springer-Verlag, New York, 1986).
[2] B. Fuchssteiner, Prog. Theo. Phys. **70** (1983) 1508.
[3] W. Strampp and W. Oevel, Prog. Theo. Phys. **74** (1985) 922.
[4] A. S. Fokas and B. Fuchssteiner, Phys. Lett. **86A** (1981) 341.
[5] A. S. Fokas, Stud. Appl. Math. **77** (1987) 253.
[6] H. H. Chen, Y. C. Lee and J. E. Lin, Phys. **90** (1983) 439.
[7] Yi Cheng, Yi-Shen Li and R. K. Bullough, J. Phys. **A21** (1988)L433.
[8] Tu Gui-Zhang, J. Phys. **A21** (1988) 1951.
[9] Tian Chou, J. Phys. **A20** (1987) 367.
[10] Tian Chou and Yi-Shen Li, J. Phys. **A20** (1987) 359.
[11] K. M. Tamizhmani, Geometrical, Group Theoretical and Singularity Structure Aspects of Certain Nonlinear Partial Differential Equations, Ph.D. Thesis, University of Madras (1986),unpublished.

Symmetries and Constants of Motion of Integrable Systems

Li Yi-Shen

Department of Mathematics, University of Science and Technology of China, Hefei 230026, China

In this paper, we give a summary of the results of our group concerning the symmetries and constants of motion and relevant algebraic structure for the "s-integrable" and "c-integrable" equations.

1. Introduction

Symmetries and constants of motion are two important concepts in many physical theories. For a classical Hamiltonian system with n-degrees of freedom, the existence of n-constants of motion in involution implies the complete integrability of the Hamiltonian system. These constants of motion can be constructed from the symmetries of the Lagrangian through Noether's theorem [1].

In the last two decades, there are many approaches [2-12] which had been used to discuss the symmetries and the conserved quantities of partial differential equations. Symmetries play a very important role in deciding the integrability of a given partial differential equation and the relation connecting the two concepts, "integrability" and "symmetry," has become a more interesting problem now. In ref. [13] the authors divided the integrable nonlinear partial differential equations into two kinds: "s-integrable" is defined as those nonlinear equations which can be linearized via the spectral transform technique and the "c-integrable" ones for nonlinear equations which can be linearized by an appropriate change of dependent variables. We shall find many interesting algebraic properties of these equations arising from their nature. In this paper, we shall give a summary of the results concerning the symmetries and constants of motion and the relevant algebraic structures of these integrable equations in 1+1 dimension.

This paper is divided into two parts. In the first part, we consider the symmetries and constants of motion and the relevant algebraic structure for "s-integrable" equations. The main idea is to find two sets of symmetries from the isospectral and nonisospectral flows by the recursion operator φ associated with the spectral problem.

Research Reports in Physics Symmetries and Singularity Structures
Editors: M. Lakshmanan · M. Daniel

In the second part, the algebraic structure for the "c-integrable equations" is considered. The main idea is to find the algebraic properties of the "c-integrable" equations from the isomorphism of the algebraic properties of the relevant linearized equations using the transformation operator.

2. Some Algebraic Properties of the "s-integrable" Equations: KdV Hierarchy

Let us begin with the typical example of the KdV hierarchy. According to the scheme proposed in ref. [14], the eigenvalue problem is given by

$$\begin{cases} \varphi_{xx} = (\lambda - u)\varphi \\ \varphi_t = A\varphi + B\varphi_x \qquad A = -\frac{1}{2} B_x . \end{cases} \tag{2.1}$$

By the condition $\varphi_{xxt} = \varphi_{txx}$, it yields

$$\lambda_t - u_t = -\frac{1}{2}B_{xxx} - 2uB_x - u_xB + 2\lambda B_x , \tag{2.2}$$

where

$$\lambda_t = \sum_{j=0}^{N} k_j(t)\lambda^{N-j} , \tag{2.3}$$

$$B = \sum_{j=0}^{N} b_j\lambda^{N-j} , \quad b_0 = \beta_0(t), \quad b_{j+1} = \frac{1}{4} \Phi b_j + \beta_j(t), \; j=0,\dots,N-1. \tag{2.4}$$

The recursion operator Φ is defined as

$$\Phi = D^2 + 4u + 2u_xD^{-1} , \quad D = \frac{\partial}{\partial x} , \quad DD^{-1} = D^{-1}D = I . \tag{2.5}$$

We get the K-dV hierarchy from (2.2),

$$u_t = 2 \sum_{i=0}^{N} (\tfrac{1}{4})^i k_{N-i}(t)\Phi^i(\tfrac{1}{2}) + \sum_{i=0}^{N} (\tfrac{1}{4})^i \beta_{N-i}(t)\Phi^i u_x . \tag{2.6}$$

The hierarchy (2.6) is associated with the isospectral or nonisospectral flow according to $\lambda_t = 0$ or $\lambda_t \neq 0$ respectively.

The isospectral flow hierarchy

$$u_t = k_\ell \equiv \Phi^\ell k_0 , \quad k_0 = u_x , \quad \ell = 0,1,2,\dots \tag{2.7}$$

has infinite number of higher symmetries (or generalized symmetries)

k_m and nonlocal symmetries τ_m^ℓ which contain independent variables x and t,

$$k_m = \Phi^m k_0 , \quad \tau_m^\ell = \Phi^m \tau_0^\ell , \quad \tau_0^\ell = (2\ell + 1)tk_{\ell-1} + \frac{1}{2} , \tag{2.8}$$

where k_m are isospectral flows with $\lambda_t = 0$ (or $k_j(t) = 0$) and $(\frac{1}{4})^m \beta_{N-m}(t) = 1$, $\beta_{N-i} = 0$, $i \neq m$ in (2.6). τ_m^ℓ are nonisospectral flows with $N = n+m-1$, $\beta_n(t) = 4^{m+n-1}(2m+1)t$ and $\beta_j = 0$, $j \neq 0$, in (2.6).

In ref. [15], we have proved that the hierarchy (2.6) possesses two sets of symmetries k_m and τ_m^ℓ , and they form an infinite dimensional Lie algebra

$$\|k_m,k_n\| = 0 , \quad \|k_m, \tau_n^\ell\| = 2(m+1)k_{m+n-1} , \|\tau_m^\ell,\tau_n^\ell\| = 2(m-n)\tau_{m+n-1}^\ell , \tag{2.9}$$

where

$$\|k_m,k_n\| = k_m'[k_n] - k_n'[k_m] , \quad k_m' k_n = \lim_{\varepsilon \to 0} \frac{\partial}{\partial \varepsilon} k_m(u + \varepsilon k_n) . \tag{2.10}$$

Similar results for the hierarchies associated with AKNS, Kaup-Newell, etc. have been proved in ref. [16-19].

It is well known [20] that the hierarchy (2.7) has infinite number of constants of motion I_n in involution,

$$\{I_m, I_n\} = \int_{-\infty}^{\infty} \frac{\delta I_m}{\delta u} D \frac{\delta I_n}{\delta u} dx = 0 .$$

I_n can be constructed from the k_n symmetries; $I_n = \int_0^1 k_n(\lambda u) u d\lambda$ and (2.7) can be written in the Hamiltonian form

$$u_t = k_n = D \frac{\delta I_n}{\delta u} . \tag{2.11}$$

It can also be given in terms of the scattering data as

$$I_n = \sum_{j=1}^{m} \frac{(2k_j)^{2n+3}}{2n + 3} + \frac{1}{4\pi i} \int_{-\infty}^{\infty} (2ik)^{2n+2} \ln|a| d\, 2ik \tag{2.12}$$

where k_j and $a(k)$ are the scattering data for the spectral problem

$$\varphi_{xx} + u\varphi = \frac{\lambda}{4}\varphi , \quad \lambda = k^2 . \tag{2.13}$$

However, the symmetries τ_m^ℓ cannot be associated with the constants of motion.

In ref.[21,22],we consider the hierarchy associated with the non-isospectral flow

$$u_t = \tau^{\ell}_{N+1} \equiv \tau \ , \ N = -1,0,1,.. \tag{2.14}$$

we have proved that (2.14) also possesses the following two sets of symmetries, $\bar{k}_m, \bar{\tau}_n$:

$$\bar{\tau}_n = \bar{\Phi}^n \bar{\tau}_0 \ , \quad \bar{\tau}_0 = \tau - k\ell \ , \quad \Phi = \Phi(1+2Nt\Phi^N)^{-\frac{1}{N}}$$

$$\bar{k}_m = \Phi^m \bar{k}_0 \ , \quad \bar{k}_0 = \begin{cases} k_0, & N = -1 \\ e^{-t}k_0 \ , & N = 0 \\ g(\Phi,t)k_0, & N \geq 1 \quad g(\Phi,t) = (1+2Nt\Phi^N)^{-\frac{1}{2N}} \end{cases} \tag{2.15}$$

They form an infinite dimensional Lie algebra

$$\|\bar{k}_m,\bar{k}_n\| = 0 \ , \ \|\bar{k}_m,\bar{\tau}_n\| = (2m+1)\bar{k}_{m+n+N}, \|\bar{\tau}_m,\bar{\tau}_n\| = 2(m-n)\bar{\tau}_{m+n+N}. \tag{2.16}$$

The hierarchy (2.14) also has infinite number of constants of motion $\bar{I}_n$ in involution

$$\{\bar{I}_m,\bar{I}_n\} = \int_{-\infty}^{\infty} \frac{\delta I_m}{\delta u} D \frac{\delta I_n}{\delta u} dx = 0 \ , \ \text{where}$$

$$\bar{I}_n = \begin{cases} e^{-(2n+3)t} I_n \ , & N = 0 \\ \sum_{\ell=0}^{\infty} h_{n+\ell N} I_{n+\ell N} \ , & N \geq 1 \ . \end{cases} \tag{2.17}$$

$$h_n = I, \ h_{n+\ell N} = \frac{-(t)^{\ell}}{\ell!}(2n+3)(2n+2N+3)\ldots(2n+2\ell N + 3)$$

$\bar{I}_n$ can also be given in terms of the scattering data

$$\bar{I}_n = \sum_{j=1}^{m} \frac{(f(k_j,t)^{2n+3}}{2n+3} + \frac{1}{i\pi} \int_{-\infty}^{\infty} (\bar{f}(ik,t)^{2n+2} \ell g| a |d\bar{f}(ik,t),$$

$$\bar{f}(\xi,t) = \xi(1 + 2tN\xi^{2N})^{-1/2N}, \ \bar{f}^2 = \bar{f}(\xi^2,t) \ \xi^3 = \lambda. \tag{2.18}$$

We note that $\bar{k}_n$ no longer yield constants of motion, for the hierarchy (2.14). Similar results for the AKNS system had been given in ref.[23]

3. The Theory [24,25]

The above structures suit not only for the KdV hierarchy, but also for more general systems, which cover many equations of physical interest. Let $u = u(x,t)$ which may be vector, belong to Schwarte space. Given the field $k_0(u)$, $\sigma_0(u)$ and an operator $\varphi = \varphi(u)$, if they satisfy the condition

i) φ is hereditary: $\varphi'[\varphi f]g - \varphi'[\varphi g]f = \varphi\{\varphi'[f]g - \varphi'[g]f\} \forall f,g$

ii) φ is strong symmetry for k_0: $\varphi'[k_0] = k_0'\varphi - \varphi k_0'$

iii) α is a parameter: $\varphi'[\sigma_0] = \sigma_0'\varphi - \varphi\sigma_0' + \alpha$

iv) β is a parameter: $\|k_0, \sigma_1\| = \beta k_0$, $\sigma_1 = \varphi\sigma_0$,

we have the following theorems:

Theorem 3.1 For the hierarchy of equations

$$u_t = k_\ell(u) \equiv \varphi^\ell k_0 \ , \quad \ell = 0,1,2,... \tag{3.1}$$

there are two sets of symmetries, i.e., k and τ symmetries given by

$$k_m = \varphi^m k_0, \quad \tau_n^\ell = \varphi^n \tau_0^\ell, \tag{3.2}$$

where $\tau_0^\ell = (\alpha\ell+\beta)tk_{\ell-1} + \sigma_0$, $n = 0,1,2,....$

An infinite dimensional Lie algebra exists for these symmetries, namely,

$$\|k_m, k_n\| = 0, \quad \|k_m, \tau_n^\ell\| = (\alpha m + \beta)k_{m+n-1}, \quad \|\tau_m^\ell, \tau_n^\ell\| = \alpha(m-n)\tau_{m+n-1}^\ell. \tag{3.3}$$

For the hierarchy of equations

$$u_t = \tau_{N+1}^\ell \equiv \tau \tag{3.4}$$

there are two sets of symmetries:

$$\bar{\bar{k}}_n = \bar{\varphi}^n \bar{k}_0, \bar{\tau}_n = \varphi^n \bar{\tau}_0, \bar{\tau}_0 = \tau - k_\ell \ , \quad \bar{\varphi} = \varphi(1 + \alpha N t \varphi^N)^{-1/N}$$

$$\bar{k}_0 = \begin{cases} k_0 \ , & N = -1 \\ e^{-\beta t} \ k_0 \ , & N = 0 \\ \bar{\varphi}_1 k_0 \ , & N \geq 1, \quad \bar{\varphi}_1 = (1 + \alpha N t \varphi^N)^{-\beta/\alpha N} \end{cases} \tag{3.5}$$

A graded infinite dimensional Lie algebra is also valid for these k_n and τ_n symmetries,

$$\|\bar{k}_m,\bar{k}_n\| = 0, \quad \|\bar{k}_m,\bar{\tau}_n\| = (\alpha m+\beta)\bar{k}_{m+n+N} \|\bar{\tau}_m,\bar{\tau}_n\| = \alpha(m-n)\bar{\tau}_{m+n+N}. \tag{3.6}$$

If the hierarchy (3.1) is a Hamiltonian, i.e.,

$$u_t = k_\ell(u) = \theta \frac{\delta I_\ell}{\delta u}, \quad I_\ell = \int_0^1 k_\ell(\lambda u)u d\lambda , \tag{3.7}$$

where θ is a Hamiltonian operator, i.e., θ is invertible anti-symmetry operator and satisfies the Jacobi identity and the parameter h is defined as

$$J'[\sigma_1] + J\sigma_1' + (\sigma_1')^*J = \alpha h, \quad J = \theta^{-1} . \tag{3.8}$$

Theorem 3.2 The hierarchy (3.4) also has infinite number of involutive constants of motion $\bar{I}_n$,

$$\{\bar{I}_m,\bar{I}_n\} = \int_{-\infty}^{\infty} \frac{\delta I_m}{\delta u}\theta \frac{\delta I_n}{\delta u} dx = 0 , \quad \text{where}$$

$$\bar{I}_n = \begin{cases} \sum_{k=0}^{n} c_k^n(-t)^{n-k} I_n , & N = -1, \beta = 0, \alpha = 1, h = 1 \\ e^{-(\alpha n+\alpha h+\beta)t} I_n , & N = 0 \\ \sum_{\ell=0}^{\infty} g_n + \ell N \, I_{n+\ell N} , & N \geq 1 \end{cases}$$

$$g_n = 1, \quad g_{n+\ell N} = \frac{(-t)^\ell}{\ell!} \Pi_{j=1}^{i}(\alpha n + \alpha j N + \alpha h + \beta) \tag{3.9}$$

Many more examples are listed in refs.[24,25]. The above theory can be used not only for continuous systems, but also for discrete systems such as Toda lattice.

Remarks

1) We have proved that if two hierarchies are gauge equivalent, then the algebraic structure of their symmetries is isomorphic [26].
2) A new approach is described to construct integrable hierarchies, their Lax operators and the relevant algebraic structure for the given nonlinear integrable equation in 1+2 dimensions, a synthetical paper is given in ref.[27].

4. Lie Algebraic Structure of G.J and Its Gauge Equivalent Yang Hierarchy [28]

The theory we mentioned in Sect. 3 is dependent on the recursion operator Φ being hereditary. But we find recently that the recursion operator Φ associated with G.J and its gauge equivalent Yang hierarchy are no longer hereditary. A new approach is therefore needed to discuss the algebraic properties of them.

Consider the generalized AKNS eigenvalue problem

$$\varphi_x = M\varphi , \quad \varphi = \begin{pmatrix} \varphi_1 \\ \varphi_2 \end{pmatrix}, \quad M = \lambda V_0 + V_1, \quad V_0 = \begin{pmatrix} -1 & 0 \\ 0 & 1 \end{pmatrix},$$

$$V_1 = \begin{pmatrix} w & u \\ v & -w \end{pmatrix} \tag{4.1}$$

which was proposed by Giachetti and Johnson [29], and is called the G.J. equation for short. Its gauge equivalent Yang equation [30] reads

$$\psi_x = \overline{M}\psi , \quad \psi = \begin{pmatrix} \psi_1 \\ \psi_2 \end{pmatrix}, \quad \overline{M} = \begin{pmatrix} s & \xi+q+r \\ -\xi-q+r & -s \end{pmatrix} \tag{4.2}$$

They are connected by the gauge transformation, $\psi_1 = \varphi_1 + \varphi_2$, $\psi_2 = i\varphi_1 - i\varphi_2$, where λ and ξ are eigenparameters, u, v, w, q, r and s are functions of x and t which belong to Schwartz space and $\xi = i\lambda$, $iq = w$, $r = i(u - v)/2$, $s = (u + v)/2$. Hierarchies of nonlinear evolution equations deduced from the isospectral eigenvalue problems of (4.1) and (4.2) respectively are given in ref. [31], which read

$$p_t = \Theta_\ell \varphi^{*\alpha+1} f_0 , \quad f_0^T = (0,0,-2), \quad \alpha = 0,1,2\dots \quad \ell = \text{parameter} \tag{4.3}$$

$$\overline{p}_t = \overline{\Theta}_\ell \overline{\varphi}^{*\alpha+1} \overline{f}_0 , \quad f_0^T = (-2i,0,0) , \tag{4.4}$$

where

$$p^T = (u,v,w), \quad \overline{p}^T = (q,r,s), \quad \overline{p} = Tp$$

$$T = \begin{pmatrix} 0 & 0 & -i \\ \frac{i}{2} & -\frac{i}{2} & 0 \\ \frac{1}{2} & \frac{1}{2} & 0 \end{pmatrix}, \quad \Theta_\ell = \begin{pmatrix} 0 & -2 & \ell u \\ 2 & 0 & -\ell v \\ -\ell u & \ell v & \ell D \end{pmatrix},$$

$$\overline{\Theta}_\ell = \begin{pmatrix} -D & -\ell s & \ell r \\ \ell s & v & -i \\ -\ell r & i & 0 \end{pmatrix}, \quad \varphi^* = \begin{pmatrix} \frac{1}{2}D+w & 0 & -\frac{v}{2} \\ 0 & -\frac{1}{2}D+w & -\frac{u}{2} \\ D^{-1}u(D+2w) & D^{-1}v(D-2w) & 0 \end{pmatrix},$$

$$\overline{\varphi}^* = \begin{pmatrix} 0 & D^{-1}(2sq - rD) & D^{-1}(-2qr - sD) \\ -r & -q & D/2 \\ -s & -D/2 & -q \end{pmatrix},$$

$$D^{-}D = DD^{-1} = I, \quad D = \frac{\partial}{\partial x}, \tag{4.5}$$

(*denotes the conjugate of the operator φ or the matrix T).

Equations (4.3) and (4.4) are Hamiltonian systems which possess infinite number of constants of motion in involution such that they are integrable Hamiltonian systems, but $\Theta_\ell \varphi^* \neq \varphi\Theta_\ell$ and $\overline{\Theta}_\ell\overline{\varphi} \neq \overline{\varphi\Theta}_\ell$. In ref.[32], we have proved that operators φ, $\overline{\varphi}$ are not hereditary. An approach which makes no use of hereditary property is used in the above reference to prove that there are also two sets of infinite number of symmetries which connect the isospectral and nonisospectral hierarchy of the eigenvalue problem (4.1) and (4.2) for the system (4.3) and (4.4) respectively and the relevant Lie algebraic structure is given.

We denote by $[M,N] = MN - NM$, the commutator of 2×2 matrix M,N and $\mathrm{Tr}M = \mathrm{Tr}N = 0$ and the Lie bracket by $\|M,N\| = M'[N] - N'[M]$, where

$$M'[N] = \lim_{\varepsilon\to 0} \frac{\partial}{\partial\varepsilon} M[V_1 + \varepsilon N] \tag{4.6}$$

Given a matrix $M = M_{ij}, (i,j = 1,2)$, we denote by M_D and M_F, its diagonal and off diagonal part respectively. We define the 2×2 matrix F_m^α as

$$F_m^\alpha \equiv [V_0, N_{m+1}^\alpha] + \ell[V_{1F}, N_{m+1,F}^\alpha] - \ell[V_1, N_{m+1,D}^\alpha], \quad m = 0,1,2\ldots \tag{4.7}$$

where

$$\alpha = 0 \text{ or } 1, \quad N_0^0 = V_0, \quad N_0^1 = xV_0$$

$$a\delta_{j,0}V_0 - N_{j,x}^\alpha + [V_0, N_{j+1}^\alpha] + [V_1, N_j^\alpha] = 0, \quad j = 0,1,2,\ldots \tag{4.8}$$

To fix the solution N_j^α, we choose the integral constants of N_D to be zero. For example

$$N_1^\alpha = x^\alpha V_{1,F}, \quad N_2^0 = \begin{pmatrix} \frac{1}{2}uv & -\frac{1}{2}(u_x - 2wu) \\ \frac{1}{2}(v_x+2wv) & -\frac{1}{2}uv \end{pmatrix},$$

$$N_2^1 = \begin{pmatrix} \frac{1}{2}(xuv) + \frac{1}{2}D^{-1}(uv) & -\frac{1}{2}(u+xu_x) + xwu \\ \frac{1}{2}(v+xv_x) + xwv & -\frac{1}{2}(xuv) - \frac{1}{2}D^{-1}(uv) \end{pmatrix}$$

Now the system (4.3) can be written as

$$V_{1,t} = F_\alpha^0, \quad \alpha = 0,1,2\ldots \tag{4.9}$$

which we call the isospectral hierarchies of (4.1) and the noniso-spectral hierarchies of (4.1) read

$$V_{1t} = F_\alpha^1, \quad \alpha = 0,1,2,\ldots \tag{4.10}$$

We have proved that

Theorem 4.1

$$\|F_m^0, F_n^0\| = 0, \quad \|F_m^1, F_n^1\| = (m-n)F_{m+n-1}^1, \quad \|F_m^0, F_n^1\| = mF_{m+n-1}^0. \tag{4.11}$$

We define the vectors

$$k_n(p) = ((F_n^0)_{12}, (F_n^0)_{21}, (F_n^0)_{11})^T, \quad \sigma_n(p) = ((F_n^1)_{12}, (F_n^1)_{21}, (F_n^1)_{11})^T \tag{4.12}$$

and the Lie bracket for the vectors k, σ, $\|k,\sigma\| = k'[\sigma] - \sigma'[k]$ where $k'[\sigma] = \lim_{\varepsilon\to 0} \frac{\partial}{\partial\varepsilon} k(p+\varepsilon\sigma)$. Then the relation (4.11) can be read as

$$\|k_m, k_n\| = 0, \quad \|\sigma_m, \sigma_n\| = (m-n)\sigma_{m+n-1}, \quad \|k_m, \sigma_n\| = mk_{m+n-1} \tag{4.13}$$

Now we rewirte the system (4.3) (or 4.9) as

$$p_t = k_\alpha(p) \quad \alpha = 0,1,2,\ldots \tag{4.14}$$

and define $\tau_n^\alpha = t\alpha k_{n+\alpha-1} + \sigma_n$, and hence we have the following theorem.

Theorem 4.2 The G.J. hierarchies (4.3) have two sets of symmetries namely k_m, τ_n^α, which form an infinite dimensional Lie algebra

$$\|k_m, k_n\| = 0, \quad \|k_m, \tau_n^\alpha\| = mk_{m+n-1}, \quad \|\tau_m^\alpha, \tau_n^\alpha\| = (m-n)\tau_{m+n-1}^\alpha. \tag{4.15}$$

We define the vectors $\bar{K}_m(\bar{p}) = Tk_n(p)$ and $\bar{\tau}_n^\alpha(\bar{p}) = T\tau_n^\alpha(p)$. The Yang hierarchies (4.4) read

$$\bar{p}_t = \bar{K}_\alpha(\bar{p}) \ , \quad \alpha = 0,1,2... \tag{4.16}$$

From the relation $\|\bar{K},\bar{\tau}\|_{\bar{p}} = T\|\bar{K},\bar{\tau}\|$, where $\|\bar{K},\bar{\tau}\|_{\bar{p}}[\bar{\tau}] = \bar{K}'_{\bar{p}}[\bar{\tau}] - \bar{\tau}'_{\bar{p}}[\bar{k}]$, $\bar{K}'[\bar{\tau}] = \lim\limits_{\varepsilon\to 0} \frac{\partial}{\partial\varepsilon} \bar{K}(\bar{p} + \varepsilon\bar{\tau})$.

Thus we have the following theorem.

Theorem 4.3 The Yang hierarchies (4.4) have two sets of symmetries, $\bar{K}_m, \bar{\tau}_n^\alpha$, and form an infinite dimensional Lie algebra

$$\|\bar{K}_m,\bar{K}_n\|_{\bar{p}} = 0 \ , \quad \|\bar{K}_m,\bar{\tau}_n^\alpha\|_{\bar{p}} = m\bar{K}_{m+n-1} \ , \quad \|\bar{\tau}_m^\alpha, \bar{\tau}_n^\alpha\|_{\bar{p}} = (m-n)\bar{\tau}_{m+n-1}^\alpha . \tag{4.17}$$

5. Some Algebraic Properties of c-integrable Equations: The Burgers Equation [33]

It is well known that the Burgers equation

$$u_t = u_{xx} + 2uu_x \tag{5.1}$$

can be transformed into the linear equation

$$s_t = s_{xx} \tag{5.2}$$

using the Cole-Hopf transformation

$$B(s,u) = 0 : \quad s - e^{D^{-1}u} = 0 . \tag{5.3}$$

It is easy to verify that the operators

$$\psi_1 = D \quad \text{and} \quad \psi_2 = 2tD + x \tag{5.4}$$

map the solution of (5.2) to itself. Operators ψ_1 and ψ_2 are strong symmetries for (5.2) and are hereditary operators. Obviously the solution s is a symmetry of itself such that (5.2) possesses two sets of symmetries with two parameters m and n

$$\bar{K}_{m,n} = \psi_1^n\psi_2^m s \ , \quad \bar{f}_{n,m} = \psi_1^n\psi_2^m f \tag{5.5}$$

where f is any solution of (5.2). Using the relation

$$\psi_2^n\psi_1^n = \sum_{k=0}^{\infty} (-1)^k k! c_n^k c_m^k \ \psi_1^{m-k} \ \psi_2^{n-k} \ , \quad c_n^k = \frac{n!}{k!(n-k)!} \ , \quad c_n^k = 0, \ k > n$$

We obtain the Lie algebra of these two sets of symmetries

$$\begin{cases} \| \bar{k}_{n,m}, \bar{k}_{i,j} \| = \sum_{k=0}^{\infty} (-1)^k k! [c_m^k c_i^k - c_j^k c_n^k] \bar{k}_{n+i-k,m+j-k} \\ \| \bar{k}_{n,m}, \bar{f}_{i,j} \| = \sum_{k=0}^{\infty} (-1)^k k! c_m^k c_j^k \bar{f}_{n+i-k,m+j-k} \\ \| \bar{f}_{n,m}, \bar{f}_{i,j} \| = 0 . \end{cases} \tag{5.6}$$

From the transformation (5.3) we get the transformation operator

$$T = B^{-1} B_u = -e^{D^{-1}u} D^{-1} , \quad T^{-1} = -De^{-D^{-1}u} \tag{5.7}$$

and upon using them, we get two strong and hereditary symmetries for the Burgers equation

$$\begin{cases} \varphi = T^{-1} \psi_1 T = D + u + u_x D^{-1} \\ \psi = T^{-1} \psi_2 T = 2t\varphi + x + D^{-1} \end{cases} \tag{5.8}$$

Theorem 5.1

Burgers equation has the following two sets of symmetries

$$k_{n,m} = -T^{-1} \bar{k}_{n,m} = \varphi^n \psi^n \sigma , \quad f_{nm} = -T^{-1} \bar{f}_{n,m} = \varphi^n \psi^m (f_x - uf) e^{-D^{-1}n} \tag{5.9}$$

which form the Lie algebra

$$\| k_{m,n}, k_{i,j} \| = \sum_{k=0}^{\infty} (-1)^k k! [c_m^k c_i^k - c_j^k c_n^k] k_{n+i-k,m+j-k}$$

$$\| k_{m,n}, f_{i,j} \| = \sum_{k=0}^{\infty} (-1)^k k! c_m^k c_i^k f_{n+i-k,m+j-k}$$

$$\| f_{m,n}, f_{i,j} \| = 0 . \tag{5.10}$$

Here we have defined

$$\varphi_{\tilde{0}} = u_x , \quad \psi_{\tilde{0}} = 2tu_x + 1$$

The four sets of symmetries in ref.[34] are subset of (5.9). In a recent paper [35], it has also been reported that the new set of symmetries are also subset of (5.9) for $f_{0n} = \psi^n(\varepsilon - u)e^{-D^{-1}u+\varepsilon x+\varepsilon^2 t}$, where $f = e^{\varepsilon x+\varepsilon^2 t}$.

It is well known that (5.1) is not a Hamiltonian system. If we consider the third order Burgers equation

$$u_t = \varphi^3\tilde{0} = \varphi^2 u_x = u_{xxx} + 3uu_{xx} + 3u^2u_x + 3u_x^2 \tag{5.11}$$

and the transformation (5.3), then (5.11) is transformed into

$$s_t = s_{xxx} \tag{5.12}$$

Using the same method described above, we obtain the following two sets of symmetries for (5.11)

$$\tilde{k}_{n,m} = \varphi^n\tilde{\psi}^m\tilde{0} \quad , \quad \tilde{f}_{n,m} = \varphi^n\tilde{\psi}^m(f_x - uf)e^{-D^{-1}u} \tag{5.13}$$

where $\tilde{\psi} = 3t\varphi^2 + x + D^{-1}$. Equation (5.12) can be written in the Hamiltonian form

$$s_t = D\frac{\delta H}{\delta s}, \text{ where } H = -\frac{1}{2}\int_{-\infty}^{\infty} s_x^2 dx$$ which possesses infinite number of constants of motion

$$I_m(s) = \int_{-\infty}^{\infty} H_m^2(s)ds$$

where

$$H_m(s) = \frac{\partial^m s}{\partial x^m}, \quad m = 1,2,\ldots.$$

The third order Burgers equation can also be written in the Hamiltonian form

$$u_t = \Theta\frac{\delta\bar{H}}{\delta u}, \text{ where } \Theta = T^{-1}D(T^{-1})^* \text{ and } \bar{H}(u) = H(e^{-D^{-1}u})$$ which also possesses infinite number of constants of motion

$$\bar{I}_m(u) = I_m(e^{-D^{-1}u}).$$

Using the same approach, we have further considered the algebraic properties of the Calogero equation [36,33]

$$v_t = v_{xxx} + 3v^2v_{xx} + 3v^4v_x + 9vv_x^2 \tag{5.14}$$

and Eckhaus-Kundu equation [37,38]

$$i\psi_t + \psi_{xx} + 2|\psi|^2_x + |\psi|^4\psi = 0 \tag{5.15}$$

as they can be transformed to the linear equations

$$s_t = s_{xxx} \tag{5.16}$$

and

$$i\varphi_t = \varphi_{xx} \tag{5.17}$$

using the transformation $s - ve^{D^{-1}v^2} = 0$ and $\varphi = \psi[D^{-1}(2|\psi|^2)]^{-\frac{1}{2}}$ respectively.

6. Thomas equation [39,38]

$$u_{xt} + u_x + u_t + u_x u_t = 0 \tag{6.1}$$

can be linearized by the transformation [40]

$$u = \lg v \quad \text{or } v = e^u \tag{6.2}$$

into

$$v_{xt} + v_x + v_t = 0 \,. \tag{6.3}$$

There are three strong and hereditary operators for (6.3)

$$\psi_1 = D_x \,, \quad \psi_2 = D_t \,, \quad \psi_3 = t(D_t + 1) - x(D_x + 1). \tag{6.4}$$

By the same approach, we find the following two sets of symmetries with three parameters m,n, ℓ, for (6.2):

$$\bar{K}_{m,n,\ell} = \psi_3^m\psi_2^n\psi_1^\ell v \,, \qquad \bar{\tau}_{m,n,\ell} = \psi_3^m\psi_2^n\psi_1^\ell f \tag{6.5}$$

where v is a solution of (6.3) and f is any solution of (6.4). They form the algebraic structure as follows:

$$\|\bar{K}_{m,n,\ell}, \bar{K}_{i,j,k}\|$$

$$= \sum_{r_1=0}^{\ell}\sum_{\ell_1=0}^{\ell-r_1}\sum_{r_2=0}^{n}\sum_{\ell_2=0}^{n-r_2}(-1)^{\ell-r_1}C^\ell_{r_1}g^{r_1,\ell_1}_{\ell,i}C^n_{r_2}g^{r_2\ell_2}_{n,r_1}\bar{K}_{m+r_2,n-\ell_2+j,k+\ell_1}$$

$$-\sum_{r_1=0}^{k}\sum_{\ell_1=0}^{k-r_1}\sum_{r_2=0}^{j}\sum_{\ell_2=0}^{j-r_2}(-1)^{k-r_1}C^k_{r_1}g^{r_1,\ell_1}_{k,m}C^j_{r_2}g^{r_2,\ell_2}_{j,r_1}\bar{K}_{i+r_2,j-\ell_2+n,\ell+\ell_1}$$

$$\|\bar{K}_{m,n,\ell}, \bar{\tau}_{i,j,k}\| =$$

$$\sum_{r_1=0}^{\ell}\sum_{\ell_1=0}^{\ell-r_1}\sum_{r_2=0}^{n}\sum_{\ell_2=0}^{n-r_2}(-1)^{\ell-r_1}C_{r_1}^{\ell}g_{\ell,j}^{r_1,\ell_1}C_{r_2}^{n}g_{n,r_1}^{r_2,\ell_2}\bar{f}_{m+r_2,n-\ell_2+j,k+\ell_1}$$

$$\| \bar{f}_{m,n,\ell}, \bar{f}_{i,j,k} \| = 0 \qquad (6.6)$$

where

$$g_{m,n}^{n,0} = 1, \quad g_{m,n}^{n-1,0} = g_{m,n}^{n-1,1} = m, \quad g_{m,n}^{r,0} = m g_{m,n-1}^{r,0} \quad m>n, r=0,1\ldots n-2$$

$$g_{m,n}^{r,\ell} = (m-\ell+1)g_{m,n-1}^{r,\ell-1} + g_{m,n-1}^{r,\ell} \quad m > n, \ r+\ell< n, \ r=0,1,\ldots,n-2$$

$$g_{m,n}^{r,\ell} = (m-\ell+1)g_{m,n-1}^{r,\ell-1}, \quad r+\ell = n .$$

Three strong and hereditary operators for (6.1) read

$$\varphi_1 = D_x + u_x, \quad \varphi_2 = D_t+u_t, \quad \varphi_3 = t(D_t+u_t+1) - x(D_x+u_x+1). \qquad (6.7)$$

Two sets of symmetries for (6.7) read

$$k_{m,n,\ell} = T\bar{k}_{m,n,\ell}, \quad f_{m,n,\ell} = T\bar{f}_{m,n,\ell}, \quad T = 1/v . \qquad (6.8)$$

They form the algebraic structure just the same as (6.6).

Acknowledgments

The author would like to thank Prof. M. Lakshmanan for inviting him to take part in the workshop on symmetries and singularity structure aspects of nonlinear dynamical systems and also for the hospitality provided to him during his visit to the department. This work was partially supported by the National Science Foundation of China and the Fund of Ministry of Education of People's Republic of China.

References

[1] V. I. Arnold, **Mathematical Methods of Classical Mechanics** (Springer-Verlag, Heidelberg, 1976).
[2] N. H. Ibragimov, **Transformation Groups Applied to Mathematical Physics** (Reidel, Boston, 1985).
[3] P. J. Olver, **Application of Lie Groups to Differential Equations** (Springer-Verlag, Heidelberg, 1986)
[4] A. M. Vinogradov, Acta Appl. Maths. **15** (1989) 3-21,161-209.
[5] H. H. Chen, Y.C.Lee, J. E. Lin, in **Advance in Nonlinear Waves** (Ed. L. Debnath) (1982), pp. 233-240.
[6] B. Fuchssteiner and A. S. Fokas, Physica **4D** (1981) 47.
[7] H. Flashka, A. C. Newell and T. Ratiu, Physica **9D** (1983) 300.
[8] W. Strampp and W. Oevel, Prog. Theo. Phys. **74** (1985) 922.
[9] A. Yu Orlov, E. I. Schulman, Theor. Mat. Fiz. **64** (1985) 323.

[10] W. Oevel, **Topics in Soliton Theory and Exactly Solvable Nonlinear Equations** (Eds. M. J. Ablowitz, B. Fuchssteiner and M. Kruskal) (World Scientific, 1987) 108.
[11] A. S. Fokas, Stud. Appl. Maths. **77** (1987) 253.
[12] R. Sarathy and K. M. Tamizhmani, "Master Symmetries of Certain Nonlinear pdes," in this volume.
[13] F. Calogero and W. Eckhaus, Inverse Problems **3** (1987) 229.
[14] Li Yi-Shen, Scientia Scinica (A) **25** (1982) 911.
[15] Li Yi-Shen and Zhu Guo Cheng, Scientia Scinica(A) **30** (1987)1243.
[16] Li Yi-Shen and Zhu Guo Cheng, J. Phys. **A19** (1986) 3713.
[17] Zhu Guo Cheng and Li Yi-Shen, Kexue Tongbao **32** (1987) 289.
[18] Zhu Guo Cheng and Li Yi-Shen, J. China Univ. of Sci. Tech. **16** (1986) 1.
[19] Zhu Guo Cheng and Ji Xiaoda, Chinese Science Bulletin **34** (1989)881.
[20] A. C. Newell, **Solitons in Mathematics and Physics** (SIAM, Philadelphia, 1985).
[21] Li Yi-Shen and Cheng Yi, Scientia Sinica (A) **31** (1988) 769.
[22] Li Yi-shen and Zhu Guo Cheng,Kexue Tongbao **32** (1987) 1308.
[23] Cheng Yi and Li Yi-Shen, J. Phys. **A20** (1987) 1951.
[24] Li Yi-shen and Cheng Yi, **Topics in Soliton Theory and Exactly Solvable Nonlinear Equations** (World Scientific, Singapore, 1987), p. 216.
[25] Li Yi-Shen and Cheng Yi, "Infinite dimensional Lie algebra and a theory on the symmetries and constants of motion in 1+1 dimensional dynamical system," ICTP Preprint SMR 185/6.
[26] Li Yi-Shen, Partial Diff. Eqns. **23** (1989) 139.
[27] Cheng Yi, "2+1 dimensional integrable hierarchies, Lax operators and relevant algebraic structures" in **Nonlinear Physics** (Ed. Gu Chashas) (Springer-Verlag, Heidelberg, 1990).
[28] Li Yi Shen, Cheng Yi, Zeng Yunbo, "A Lie algebraic structure of G.J. and its gauge equivalent Yang hierarchies," in **Nonlineary Physics** (Ed. Gu Chaohao)
[29] R. Giachettiand R.Johnson, Physica **102A** (1984) 81.
[30] C. N. Yang, Commun. Math. Phys. **112** (1987) 205.
[31] Li-Yi Shen, Chinese Sci. Bull. **34** (1989) 567.
[32] Li-Yi Shen and Zeng Yunbo, "On some properties of G.J. equation and its gauge equivalent Yang equation" (to be published).
[33] Li Yi-Shen, "Some algebraic properties of the c-integrable nonlinear equation I-Burgers equation and Calogero equation," Scientia Scinica **A** (1990), to appear.
[34] Tian Chow, Scientia Scinica **A31** (1988) 141.
[35] Tao Sun, J. Phys. **A22** (1989) 3737.
[36] F. Calogero, J. Math. Phys. **28** (1987) 538.
[37] A. Kundu, Physica **25D** (1987) 399.
[38] Li Yi-Shen, "Some algebraic properties of c-integrable nonlinear equation II-Echkahus-Kundu equation and Thomas equation" Preprint.
[39] H. C. Thomas, J. Am. Chem. Soc. **66** (1944) 1664.
[40] R. R. Rosales, Stud. Appl. Math. **59** (1978) 117.

Lie Algebra, Bi-Hamiltonian Structure and Reduction Problem for Integrable Nonlinear Systems

S. Roy and A. Roy Chowdhury

High Energy Physics Division, Department of Physics,
Jadavpur University, Calcutta 700032, India

Hamiltonian structures of integrable nonlinear evolution equations are studied in the framework of infinite dimensional Lie Algebra. We have shown that it is actually possible to derive the two-symplectic structures from the formalism of G. Zhang Tu. Furthermore we have extended this formalism to a system with 3×3 matrix structure. An important aspect of our formulation is to implement reduction mechanism to arrive at a specific nonlinear system. As examples we have discussed the cases of KdV, Langmuir solitons.

1. Introduction

Integrable nonlinear systems occupy a central place in the present-day study of dynamical systems. The emergence of soliton as a potentially viable physical object in many branches of physics is solely responsible for such a phenomenon [1]. With the initial impetus gained by the astounding discovery of Lax [2], there has been tremendous amount of research in the field, leading to concepts such as complete integrability as per Painlevé [3], prolongation structure as per Wahlquist-Estabrook [4], Lie-Bäcklund symmetries, hereditary or master symmetries [5] and so on. As with any dynamical problem, it is a quite natural attempt to understand such a problem in the domain of Hamiltonian structure [6]. And in the course of such developments, people have discovered the important and novel aspect of bi-Hamiltonian structure associated with such nonlinear systems [7]. It was really an important observation that the recursion operator for the Lie-Bäcklund symmetries really factorizes into two symplectic operators. Of late there have been many attempts to deduce the bi-Hamiltonian structure via actual LB symmetry calculation [8], via squared eigenfunctions [9], and also via Lie algebraic ideas [10]. Recently, an elegant formalism has been laid down by G. Zhang Tu [11], using Lie algebra and a relation known as trace identity for arriving at the Hamiltonian structure of NLPDE. Here in this communication we show that it is possible to utilise this set up to deduce the bi-

Research Reports in Physics
Editors: M. Lakshmanan · M. Daniel

Symmetries and Singularity Structures

Hamiltonian structure and hence the recursion operator. Also, we have extended the formalism to Lax system with SU(3) or SL(3,R) group structure. As an example of such extension we treat the case of Langmuir solitons [12]. The usual problem associated with a Lax pair having bigger Lie group structure is the redundancy of many of the matrix elements. Whence comes in the problem of reduction elegantly formulated by Mikailov [13]. We here also demonstrate that it is possible to incorporate such a reduction mechanism in this formalism and to scoop out the particular choice of nonlinear systems, with the associated bi-Hamiltonian structure.

2. Formulation

As integrable nonlinear partial differential equation is usually associated with a Lax pair written as

$$\Psi_x = U\Psi \; ; \quad \Psi_t = V\Psi \tag{1}$$

whence U, V are functions (they can be matrices also) involving the nonlinear field variables and their derivatives. When the Lax pair is not a scalar one, our U and V will be matrices belonging to some specific Lie algebra. The central problem is to derive the Hamiltonian structure of the evolution equations when the coefficients of V are determined via

$$U_t - V_x = [U,V] \; . \tag{2}$$

Since the matrix U and V are expanded in positive and negative powers of λ, we can assert that in general they are elements of a loop algebra [14].

Let g be a loop algebra and the degree of the elements of g are defined by

$$\deg(x \otimes \lambda^n) = nx \in g \tag{3}$$

and

$$g_+ = \sum_{n \geq 0} g_n \; , \quad g_- = \sum_{n < 0} g_n . \tag{4}$$

The Killing-Cartan form can also be defined on g, depending upon the particular algebra chosen. If R and E_i are the generators of g then we set

$$U = R + \sum_{i=1}^{p} u_i E_i \; . \tag{5}$$

We note now the important trace identity derived by Zhang. If V is a solution of Eq. (2) then we can always find a $\mu > 0$, such that $\bar{V}$

$= \lambda^{\mu} V$ is also a solution of the same equation; then

$$\left(\frac{\delta}{\delta u_i}\right) \mathrm{tr}\left(\bar{V} \frac{\delta U}{\delta \lambda}\right) = \frac{\partial}{\partial \lambda} \mathrm{tr}\left(\bar{V} \frac{\partial U}{\partial u_i}\right) , \qquad i = 1,\ldots,p \tag{6}$$

or

$$\left(\frac{\delta}{\delta u_i}\right) \langle V, \frac{\partial U}{\partial \lambda} \rangle = \left(\lambda^{\mu}\frac{\partial}{\partial \lambda} \lambda^{\mu}\right) \langle V, \frac{\partial U}{\partial u_i} \rangle \tag{7}$$

where $\langle x,y\rangle = \mathrm{tr}(x,y)$.

In the sequel equation, (6) or (7) will play an important role in the derivation of the bi-Hamiltonian form. If now J and L are two skew symmetric operators then the evolution equations generated via Eq. (1) can be cast in the form

$$U_t = JL^n f(U) \tag{8}$$

where $L^n f(U) = \frac{\delta H_n}{\delta U}$ and these H_n's are considered to be the Hamiltonians with $\{H_n, H_m\} = 0$ with respect to the Poisson bracket

$$\{f,g\} = \left(\frac{\partial f}{\partial U} , J \frac{\partial g}{\partial U}\right) . \tag{9}$$

3. The AKNS Problem and KdV Equation

We start with the age-old KdV problem from the AKNS formulation as it will clarify the basic points of computation and will exhibit the relation of the reduction mechanism in the Hamiltonian formulation. The AKNS spectral problem is written as

$$\Psi_x = U\Psi = (-i\lambda\sigma_3 + q\sigma_+ + r\sigma_-)\Psi , \tag{10}$$

$\sigma_+ = \sigma_1 + i\sigma_2$, $\sigma_- = \sigma_1 - i\sigma_2$ and σ_3 are the Pauli-matrices. Let us set

$$\sigma_3 \otimes \lambda^n = \sigma_3(n), \quad \sigma_+ \otimes \lambda^n = \sigma_+(n), \quad \sigma_- \otimes \lambda^n = \sigma_-(n) . \tag{11}$$

The commutation rules are

$$[\sigma_3(n), \sigma_+(m)] = 2\sigma_+(m + n) ,$$

$$[\sigma_3(m), \sigma_-(n)] = -2\sigma_-(m + n) ,$$

$$[\sigma_+(m), \sigma_-(n)] = \sigma_3(m + n) . \tag{12}$$

We now set

$$V = \sum_j \{a_j\sigma_3(-j) + b_j\sigma_+(-j) + c_j\sigma_-(-j)\} \quad . \tag{13}$$

Evaluating

$$\langle V, \frac{\delta U}{\delta\lambda} \rangle = -2i\, a_j\, \lambda^{-j} \text{ and } \langle V, \frac{\delta U}{\delta q} \rangle = c_j\, \lambda^{-j} \, . \tag{14}$$

We now go back to Eq. (13) with V replaced by $\lambda^n V$. It is easy to ascertain that

$$\lambda^n V = \sum_j \{a_j\sigma_3(n-j) + b_j(\sigma_+(n-j) + c_j\sigma_-(n-j)\} \, , \tag{15}$$

whence we compute

$$U_t - \lambda^n V_x + [U, \lambda^n V] = 0 \, , \tag{16}$$

which yields

$$q_t\sigma_+(0) - [a_{jx}\sigma_3(n-j) + b_{jx}\sigma_+(n-j) + c_{jx}\sigma_-(n-j)] - 2ib_j\sigma_+(n-j+1)$$

$$+ 2ic_j\sigma_-(n-j+1) - 2qa_j\sigma_+(n-j) + qc_j\sigma_3(n-j) - 2ia_jr\sigma_-(n-j)$$

$$+ rib_j\sigma_3(n-j) = 0 \, . \tag{17}$$

The recursion relation for the coefficients (a_j, b_j, c_j) are

$$2ib_{j+1} = -b_{jx} - 2qa_j \, ,$$

$$2ic_{j+1} = c_{jx} + 2ra_j \, ,$$

$$a_{jx} = qc_j + rb_j \, . \tag{18}$$

To reduce to the KdV case we invoke the Z_2-reduction [15]

$$U(-\lambda)G(\lambda) = G(\lambda)U(\lambda) \, , \tag{19}$$

$$G(\lambda) = \begin{pmatrix} 1 & -2\lambda \\ 0 & 1 \end{pmatrix} \, .$$

Condition (19) immediately yields r = -1, giving the KdV hierarchy. We now impose (19) also on the time part V, that is,

$$V(-\lambda)G(\lambda) = G(\lambda)V(\lambda) \tag{20}$$

leading to the condition

$$C_{n+1} = 0 \, . \tag{21}$$

Now setting $r = -1$ in (18) and using (21) we deduce

$$-2D\, C_{j+1} = (\tfrac{1}{2} D^3 + iq_x + 2iqD)C_{j-1} \tag{22}$$

which is nothing but the Lenard relation [16]. Now the Hamiltonian set is defined via the Killing form

$$H = Tr(V, \frac{\delta U}{\delta\lambda}) \, ,$$

leading to the expressions of H in terms of the coefficients a_j and c_j. Now from the relation

$$q_t = 2ib_{n+1} = (\tfrac{1}{2} D^3 + iq_x D + 2iqD)C_n = -2DC_{n+2} \, . \tag{23}$$

Lastly from (14), (23) and (6) it is quite apparent that C_j's are nothing but the variational derivatives of the H_j's. Hence (23) is the bi-Hamiltonian structure of the KdV system. Actually we define $H = H_n \lambda^{-n}$. So from (22)

$$H_n = \frac{1}{n-1}(-2ia_n) \, . \tag{24}$$

4. Langmuir Soliton Equation and $SU_3 \otimes C(t,t^-)$ Algebra

We now extend our above discussion to a more complicated and less discussed situation of Langmuir soliton equations, for which the inverse problem is a 3×3 one. Previously the Hamiltonian structure and conservation laws were discussed by the present authors with the help of squared eigenfunctions. The algebra $SU_3 \otimes C(t,t^{-1})$ is generated by

$$\begin{aligned}
I_+ \otimes \lambda^n &= I_+(n) \, , & I_- \otimes \lambda^n &= I_-(n) \, , \\
U_- \otimes \lambda^n &= U_-(n) \, , & U_+ \otimes \lambda^n &= U_+(n) \, , \\
V_+ \otimes \lambda^n &= V_+(n) \, , & V_- \otimes \lambda^n &= V_-(n) \, , \\
I_3 \otimes \lambda^n &= I_3(n) \, , & U_3 \otimes \lambda^n &= U_3(n) \, ,
\end{aligned} \tag{25}$$

where I_+, I_-, U_+, U_-, etc. are the I-spin, U-spin generators of SU_3. The basic commutation rules are

$$[U_3(n), I_\pm(m)] = \pm \frac{1}{2} I_\pm(m+n) , \quad [V_3(n), I_\pm(m)] = \pm \frac{1}{2} I_\pm(m+n),$$

$$[I_+(n), I_-(m)] = 2I_3(m+n), \quad [V_+(n), V_-(m)] = 2V_3(m+n) ,$$

$$[V_+(m), V_+(n)] = U_-(m+n) , \text{ etc.} \tag{26}$$

The Lax pair is as follows:

$$\Psi_x = \{2i\lambda(I_3 + U_3) + AI_+ + iBV_- + A^*U_- - iD_1V_+ + D_3I_- + D_2U_-\}\Psi$$

$$= U\Psi , \tag{27}$$

and

$$\Psi_t = V\Psi ,$$

$$V = \sum\{a_jI_3(-j) + b_jU_3(-j) + c_jV_3(-j) + d_jI_+(-j)$$

$$+ e_jI_-(-j) + f_jU_+(-j) + g_jU_-(-j) + h_jV_+(-j) + k_jV_-(-j)\}. \tag{28}$$

Then the equation $U_t - \lambda^n V_x + [U, \lambda^n V] = 0$ leads to

$$A_t = - id_{n+1} , \quad -iD_{1t} = 2ih_{n+1} ,$$

$$iB_t = -2ik_{n+1} , \quad D_{2t} = ie_{n+1} ,$$

$$A^*_t = -if_{n+1} , \quad D_{3t} = ig_{n+1} . \tag{29}$$

Our space part of the Lax equation contains six nonlinear field variables which give rise to six evolution equations (29). But a physical case like Langmuir system contains only three of them, one complex and one real field. A convenient form of Langmuir equations was noted in reference [17]. We now try to make a reduction of the present system to this case.

Following Konopelochenko we consider the condition

$$U(-\lambda)\tilde{G}(\lambda) = \tilde{G}(\lambda)U(\lambda) \tag{30}$$

where this time $\tilde{G}(\lambda)$ is a 3×3 matrix written as

$$\tilde{G}(\lambda) = \begin{pmatrix} 1 & 0 & -2\lambda \\ 0 & 1 & 0 \\ 0 & 0 & 1 \end{pmatrix} . \qquad (31)$$

If Eq. (30) is explicitly evaluated it immediately yields

$$D_1 = 1, \quad D_2 = 0 , \quad D_3 = 0 . \qquad (32)$$

So from Eq. (29) we can assert

$$h_{n+1} = e_{n+1} = g_{n+1} = 0 . \qquad (33)$$

On the other hand these supplementary equations (33) can be utilised to convert a six by six recursion relation of the coefficients to a 3 × 3 one written as

$$\begin{pmatrix} e_{m+1} \\ g_{m+1} \\ ih_{m+1} \end{pmatrix} = \begin{pmatrix} D^2-B-\frac{3i}{2}A*D^{-1}A & \frac{3i}{2}A*D^{-1}A* & \frac{3i}{2}A*D+iA^*_x \\ -\frac{3i}{2}AD^{-1}A* & D^2-B+\frac{3i}{2}AD^{-1}A* & -\frac{3i}{2}AD-iA_x \\ \frac{A}{4}-\frac{1}{2}D^{-1}AD & -\frac{A*}{4}-\frac{1}{2}D^{-1}A*D & \frac{D^2}{4}-\frac{1}{2}B-\frac{1}{2}D^{-1}BD \end{pmatrix} \begin{pmatrix} e_{m-1} \\ g_{m-1} \\ ih_{m-1} \end{pmatrix}$$

On the other hand the trace identity immediately leads to the connection of e_j, g_j, ih_j with the variational derivatives of H as

$$\langle V, \frac{\delta U}{\delta A} \rangle = e_j\lambda^{-j} \;;\quad \langle V, \frac{\delta U}{\delta A*} \rangle = g_j\lambda^{-j} \;;\quad \langle V, \frac{\delta U}{\delta B} \rangle = ih_j\lambda^{-j}$$

and

$$\langle V, \frac{\delta U}{\delta \lambda} \rangle = -\frac{i}{2}(a_j + b_j + c_j)\lambda^{-j} .$$

Lastly the evolution equations obtained are

$$\begin{pmatrix} A \\ A* \\ iB \end{pmatrix} = \Omega_1 \begin{pmatrix} e_{n-2} \\ g_{n-2} \\ ih_{n-2} \end{pmatrix} = \Omega_2 \begin{pmatrix} e_n \\ g_n \\ ih_n \end{pmatrix} ,$$

where the symplectic operators Ω_1, Ω_2 are given as,

$$\Omega_1 = \begin{pmatrix} -\frac{3}{2}AD^{-1}A & -iD^2+iB+\frac{3}{2}AD^{-1}A^* & -A_x-\frac{3}{2}AD \\ iD^2-iB+\frac{3}{2}A^*D^{-1}A & -\frac{3}{2}A^*D^{-1}A^* & -A^*_x-\frac{3}{2}A^*D \\ -\frac{i}{2}A_x-\frac{3i}{2}AD & -\frac{i}{2}A^*_x-\frac{3i}{2}A^*D & \frac{i}{2}D^3-iB_x-2iBD \end{pmatrix},$$

$$\Omega_2 = \begin{pmatrix} 0 & i & 0 \\ -i & 0 & 0 \\ 0 & 0 & -2iD \end{pmatrix}.$$

So it is the same bi-Hamiltonian structure as obtained in reference [17] via the squared eigenfunction approach.

5. Discussion

In our above analysis we have shown that it is possible to utilise the trace identity of Zhang to obtain the bi-Hamiltonian structures of various classes of nonlinear systems integrable via Lax pair. The two symplectic forms have immense utility in yielding the recursion operator or hereditary operator for the symmetry. The Lie algebraic framework provides very useful steps for practical computation. The original idea of using Lie algebraic framework can be traced back to the works of Wilson, Kupershmidt, Drienfeld and Sokolov. While their formulation gives an abstract framework for the nonlinear systems, our approach yields a concrete computational methodology.

Acknowledgment

One of the authors (S.R.) is grateful to the Department of Science and Technology, Government of India for a young scientist project grant.

References

[1] R. K. Bullough and P. J. Caudrey, eds., **Solitons** (Springer-Verlag, Berlin, 1980).
[2] P. D. Lax, Commun. Pure Appl. Math. **21** (1968) 467.
[3] J. Weiss, M. Tabor and J. Carnevale, J. Math. Phys. **24** (1983) 522.
[4] H. Wahlquist and F. Estabrook, J. Math. Phys. **16** (1975) 1.
[5] B. Fuchssteiner, Prog. Theo. Phys. **68** (1982) 1082; A. Fokas and B. Fuchssteiner, Physica **D4** (1981) 47.
[6] S. Novikov, ed., **Integrable Systems**, Lecture Notes Series (Cambridge Univ. Press, London, 1981)

[7] B. Konopelochenko, Lecture Notes in Physics 270 (Springer-Verlag, Berlin, 1986).
[8] A. Fokas, J. Math. Phys. **21** (1980) 1318.
[9] A. C. Newell, in **Solitons**, ed. R. K. Bullough and P. J. Caudrey (Springer-Verlag, Berlin, 1980).
[10] G. Wilson, Ergod. Th. and Drynam. Sys. **1** (1981) 361; V. G. Drienfeld and V. V. Sokolov, Dokl. Akad. Nauk. SSSR **258** (1981); B. Kuperschmidt, Physica **D27** (1987) 294; Physica **D7** (1983) 334; Phys. Lett. **109A** (1985) 417.
[11] G. Zhang Tu, Scientia Sinica **A29** (1986) 138.
[12] N. Yajima and M. Oikawa, Prog. Theo. Phys. **56** (1976) 1719.
[13] A. Mikailov, Physica **D3** (1981) 73.
[14] G. Segal and A. Presely, **Loop Groups** (Oxford Univ. Press, London, 1986).
[15] B. G. Konopelchenko, Phys. Lett. **108** (1982) 26.
[16] G. Eilenberger, **Soliton** (Springer-Verlag, Berlin, 1981).
[17] S. Roy and A. Roy Chowdhury, J. Phys. **A21** (1988) L277.

On the Role of Virasoro, Kac–Moody Algebra and Conformal Invariance in Soliton Hierarchies

A. Roy Chowdhury

High Energy Physics Division, Department of Physics,
Jadavpur University, Calcutta 700032, India

The implications of infinite dimensional Lie algebra, Virasoro and Kac-Moody algebras in the study of completely integrable partial differential equations, are reviewed. Intimate relations are shown to exist between prolongation algebra, Lie-Bäcklund algebra and such infinite dimensional algebras. The importance of conformal invariance is emphasized and some new aspects of complete integrability are exhibited. Lastly the relation between bi-Hamiltonian structure and conformal invariance is explained.

1. Introduction

The study of completely integrable nonlinear partial differential equations (nlpde's) in two dimensions has proved to be of prime importance over the last twenty years. The discovery of Lax pair and inverse scattering gave a further boost to such studies [1]. On the other hand several authors formulated a technique to test the complete integrability which goes by the name of Painlevé analysis [2]. Also it has been proved that such nlpde's possess infinite number of conservation laws [3] whose implications are yet not clear. Since the preliminary analysis depends on a Lax pair which is defined in the setting of a Lie algebra [4] people have been trying to utilise several properties of Lie algebras to extract further information about the properties of such nlpde's. Furthermore it is seen in recent years that both the finite and infinite dimensional [5] Lie algebras can be useful. Here in this talk we want to display the elegant role played by such infinite dimensional Lie algebra in various analysis of nlpde's.

2. Infinite Dimensional Algebra and Prolongation Analysis

After the initial discovery of the Lax pair for the KdV problem it was a real problem to obtain the corresponding linear problem for other nlpde's, until the famous thesis of Wahlquist and Estabrook [6]. In this paper the authors show how a differential geometric setting can be used to deduce the Lax pair of a given nonlinear problem. Ela-

Research Reports in Physics Symmetries and Singularity Structures
Editors: M. Lakshmanan · M. Daniel

boration of this basic theory can be found in many places specially in references [6,7]. In the later years this technique has been proved to be extremely useful. The ultimate culmination of such an analysis is the derivation of a closed Lie algebra, yielding a finite dimensional form of the IST problem. Here we demonstrate that even an infinite dimensional form of the Lie algebra will give important information about the Lax pair. Let us demonstrate the whole thing on the basis of an example--the sine-Gordon theory. This equation is written as[7],

$$\varphi_{uv} = \sin\varphi \ , \tag{1}$$

with u, v as light cone coordinates. To proceed with the prolongation analysis the usual practice is to rewrite this as a pair of first order equations

$$\varphi_u = a \ ; \quad a_v = \sin\varphi \ . \tag{2}$$

Equations (2) can be converted into the differential form language as

$$F_1 = -d\varphi\wedge dv + a\,du\wedge dv$$

$$F_2 = da\wedge du + \sin\varphi\ du\wedge dv. \tag{3}$$

These are closed under exterior differentiation, so that there exists one form η_i^k such that

$$dF_i = \sum_k \eta_i^k F_k \ . \tag{4}$$

Next step is to introduce pseudopotentials y^α and one forms w^α, so that

$$w^\alpha = - dy^\alpha + A^\alpha(a,\varphi, y^\beta)du + B^\alpha(a,\varphi, y^\beta)dv \tag{5}$$

along with the condition

$$dw^\alpha = \lambda^{\alpha_i} F_i + \mu_\beta^\alpha w^\beta. \tag{6}$$

Using Eqs. (6), (5) and (3) we get

$$A\varphi = Ba = 0$$

$$[A,B] = - Aa \sin\varphi + B_\varphi a \ . \tag{7}$$

The most suitable forms of A and B are taken to be

$$A = A_1(y)a + A_2(y)$$

$$B = A_3(y)\sin\varphi + A_4(y)\cos\varphi$$

which when inserted in (7) yields

$$[A_1,A_3] = -A_4\ , \qquad [A_1,A_4] = A_3\ ,$$

$$[A_2,A_3] = -A_1\ , \qquad [A_2,A_4] = 0\ . \tag{8}$$

Thus we have four generators A_1, A_2, A_3 and A_4. Equation (8) does not yield the complete information of a Lie algebra generated by them because two more commutation rules are needed for the closure of the algebra. These are $[A_1,A_2]$ and $[A_3,A_4]$. The initial approach of ref. [6] and many other papers were to impose some relations on these A_i's so that the number of unknown commutators get reduced, and thereby yielding a finite closed Lie algebra. But instead of this standard approach we first write down all the Jacobi identities in connection with the four generators A_i ($i = 1,2,3,4$) leading to

$$[[A_1,A_2],\ A_3] = 0$$

$$[[A_3,A_4],\ A_1] = 0$$

$$[[A_1,A_2],\ A_4] = A_1$$

$$[[A_3,A_4],\ A_2] = A_3\ . \tag{9}$$

We can now follow the mechanism described above and define a new pair of generators through

$$[A_1,A_2] = A_5\ ; \qquad [A_3,A_4] = A_6\ . \tag{10}$$

We now compute many more Jacobi identities and at each stage introduce new A_i's. At this stage one may note that the commutators $[A_1,A_3]$ and $[A_1,A_4]$ are suggestive of a $O(2,1)$ structure. We therefore rename the generators as follows:

$$A_1 = X_0, \quad A_3 = Y_0\ , \quad A_4 = Z_0\ .$$

Observing that $[A_3,A_4]$ commutes with A_1, we set $A_6 = X_1$. Proceeding along these lines we end up with

$$[X_i,Y_k] = -Z_{i+k}\ ; \quad [Y_i,Z_k] = X_{i+k+1}\ ; \quad [Z_i,X_k] = -Y_{i+k}\ ;$$

$$[X_i,X_k] = [Y_i,Y_k] = [Z_i,Z_k] = 0\ . \tag{11}$$

Since the indices run over infinite values, they clearly generate an infinite dimensional Lie algebra. One can, of course, find some representation of such infinite dimensional algebras in terms of infinite matrices whose elements are $3 * 3$ matrices. One such representation can be constructed as

$$X^{\alpha}_{i\,\beta} = \delta^{\alpha-i}_{\beta}\, e$$

$$Y^{\alpha}_{i\,\beta} = \delta^{\alpha-i}_{\beta}\, \ell_1 + \delta^{\alpha-i-1}_{\beta}\, \ell_2$$

$$Z^{\alpha}_{i\,\beta} = \delta^{\alpha-i}_{\beta}\, m_1 + \delta^{\alpha-i-1}_{\beta}\, m_2 \tag{12}$$

with

$$e = \begin{pmatrix} 0 & 0 & 0 \\ 0 & 0 & 1 \\ 0 & -1 & 0 \end{pmatrix} ; \quad \ell_1 = \begin{pmatrix} 0 & 0 & 1 \\ 0 & 0 & 0 \\ 0 & 0 & 0 \end{pmatrix}$$

$$\ell_2 = \begin{pmatrix} 0 & 0 & 0 \\ 0 & 0 & 0 \\ 1 & 0 & 0 \end{pmatrix} ; \quad m_1 = \begin{pmatrix} 0 & -1 & 0 \\ 0 & 0 & 0 \\ 0 & 0 & 0 \end{pmatrix} ; \quad m_2 = \begin{pmatrix} 0 & 0 & 0 \\ -1 & 0 & 0 \\ 0 & 0 & 0 \end{pmatrix}$$

It is rather interesting to observe that our algebra is different from $O(2,1) \otimes R(t,t^{-1})$ due to the structure of the commutators in (11). It is actually a subalgebra of $SL(3,R) \otimes R(t,t^{-1})$. There are further interesting implications of these infinite dimensional Lie algebras, which are outside the scope of the present lecture, for which we refer to the original literature [8]. Instead, we now write down the explicit form of the differential one form w as

$$w = -dy_{\alpha} + [(ae + m_1)y_{\alpha} + m_2 y_{\alpha+1}]du$$
$$+ [(\sin\varphi \ell_1 + \cos\varphi m_1)y_{\alpha-1} + (\sin\varphi \ell_2 + \cos\varphi m_2)y_{\alpha}]dy. \tag{13}$$

where y_{α} are 3-dimensional vectors and α ranges from $-\infty$ to $+\infty$. Restriction to a solution manifold $w_{\alpha} = 0$ gives linear equations for ξ_{α}, whose integrability leads to the sine-Gordon problem. Assuming

$$\xi(t) = \sum_{-\infty}^{\infty} \xi_{\alpha} t^{\alpha}$$

we can get

$$\xi_u(t) = (ae + m_1 + \frac{m_2}{t})\, \xi(t) ,$$

$$\xi_v(t) = [t(\sin\varphi \ell_1 + \cos\varphi m_1) + \sin\varphi \ell_2 + \cos\varphi m_2]\xi(t), \tag{14}$$

which are the explicit forms of the Lax equations with 't' as the spectral parameter.

3. Use of Infinite Dimensional Algebra for Obtaining Bäcklund Transformation

The analysis described above can also be performed for the KdV case as done in ref. [9]. We here demonstrate that, by an explicit realization of the generators, one can construct more general Bäcklund transformations for the KdV problem. The zero curvature equation in the KdV case is written as

$$\Psi_x = U\Psi \; ; \quad \Psi_t = V\Psi \tag{15}$$

where U, V which belong to the algebra $A_1 \otimes C(t)$ are written as

$$U = y - (2u + T)z$$

$$V = (2u_{xx} + 8u^2 - 4T^2 - 4uT)z + (4T - 4u)y - 2u_x h. \tag{16}$$

The KdV equation is considered in the form

$$u_t + 12uu_x + u_{xxx} = 0 \tag{17}$$

with $[h,y] = 2y; \quad [h,z] = -2z \; ; \quad [y,z] = h$.

Let us suppose that u' is another solution of the KdV equation and we can have another Lax pair [10]

$$\Psi'_x = u'\Psi' \, , \quad \Psi'_t = v'\Psi' \, . \tag{18}$$

Now, assume ψ and ψ' to be connected via

$$\psi' = g\,\psi \tag{19}$$

where $g \in A_1$ and is parameterized as $g = e^{\alpha h}e^{\beta y}e^{\gamma z}(\alpha,\beta\;,\gamma)$ are arbitrary functions of (x,t), whence one immediately gets

$$A' = g_x g^{-1} + g\,A\,g^{-1}$$

$$B' = g_t g^{-1} + g\,B\,g^{-1} \quad . \tag{20}$$

If the explicit form of g is now used, we obtain

$$\alpha_x = \gamma - \beta(2u + T)e^{2\alpha}$$

$$\gamma_x = -(2u' + T) + (2u + T)e^{2\alpha} - \gamma^2 \tag{21}$$

$$\beta_x = 1 - e^{-2\alpha} + \beta^2(2u + T)e^{2\alpha} + 2\beta\gamma \quad .$$

A similar but somewhat complicated set of equations can be obtained for the time part. We now impose the condition that it is possible to get u from u' via g^{-1} which is g itself with $(-\alpha, -\beta, -\gamma)$. This yields

$$e^{2\alpha} = \left(\frac{2u' + T}{2u + T}\right)^{\frac{1}{2}} \tag{22}$$

$$\gamma = \int(u' - u)dx + g_1(t)$$

and from (21) we get

$$[\int(u - u')dx + g_1(t)]^2 = -a' - u - T + (2u + T)^{\frac{1}{2}}(2u'+T)^{\frac{1}{2}} \tag{23}$$

which is nothing but a new Bäcklund transformation involving two parameters $(T, g_1(t))$. One can observe that $u = u'$ if $g_1 = T = 0$. If we set $g_1 = 0$, $u' = 0$ in the above equation we get

$$v^2 = 2u + T$$

$$\frac{dv}{dx} = \frac{1}{v^2}(v - T) \tag{24}$$

So, $v = -T^{\frac{1}{2}} \tanh \frac{\alpha x+h'}{2}$ and hence

$$u = \frac{\alpha^2}{4} \operatorname{sech}^2(\frac{\alpha x+h'}{2}) \tag{25}$$

which is nothing but the one soliton solution.

4. Conformal Invariance of Nonlinear Equation and Virasoro Algebra

Now let us look at a problem slightly different from the previous one. Consider the Liouville equation in two dimensions [11], which is the source for understanding various problems in string theory and nlpde's. This equation is written as

$$\partial_{+-}\varphi = e^{\varphi} \tag{26}$$

where ∂_+, ∂_- denote derivatives with respect to light cone coordinates. The zero curvature representation for (26) is

$$\partial_+A_- - \partial_-A_+ = [A_+, A_-] \tag{27}$$

which is the result of self-consistency for

$$\partial_+\Psi = A_+\Psi \quad ; \quad \partial_-\Psi = A_-\Psi \tag{28}$$

A_+, A_- are given as

$$A_+ = u^+H + f^+E^+ = g^{-1}\partial_+ g$$

$$A_- = u^-H + f^-E^- = g^{-1}\partial_- g$$

$$g \in SL(2,c) \ . \tag{29}$$

If we invoke the gauge invariance of Eq. (28) with the gauge function belonging to SL(2,c), and parametrize g as ($\Psi' = g\Psi$),

$$g = (1 + m^+y)(1 + n^-z)e^{Hk^+}$$

$$= (1 + m^-z)(1 + n^+y)e^{Hk^-} \tag{30}$$

along with

$$e^{K^+-K^-} = (1 - m^-m^+)$$

we obtain

$$A_{\pm} = g^{-1}A_{\pm}g + g^{-1}\partial_{\pm}g$$

$$e^{\varphi'} = \frac{e^{-\varphi}\left[e^{\varphi} + \partial_+\left\{\frac{\alpha e^{\varphi}}{1 - \frac{\alpha}{2}\partial_+\varphi}\right\}\right]\left[e^{\varphi} + \partial_-\left\{\frac{\beta e^{\varphi}}{1 - \frac{\beta}{2}\partial_-\varphi}\right\}\right]}{\left\{\frac{\alpha\beta\, e^{\varphi}}{(1 - \frac{\alpha}{2}\partial_+\varphi)(1 - \frac{\beta}{2}\partial_-\varphi)}\right\}} \tag{31}$$

where α, β are arbitrary functions of x_+, x_- respectively. Equation (31) yields an auto Bäcklund transformation of the solutions. Now we choose α, β to be infinitesimal and then

$$\delta\varphi = \partial_+\alpha + \partial_-\beta + \alpha\partial_+\varphi + \beta\partial_-\varphi \ . \tag{32}$$

The generators corresponding to infinitesimal translation along α or β are

$$L = \varepsilon\frac{\partial}{\partial\alpha} = \varepsilon\left(\frac{\partial\alpha}{\partial x_+}\right)^{-1}\frac{\partial}{\partial x_+}$$

$$\bar{L} = \delta\frac{\partial}{\partial\beta} = \partial\left(\frac{\partial\beta}{\partial x_-}\right)^{-1}\frac{\partial}{\partial x_-} \ . \tag{33}$$

If we choose $\alpha = \varepsilon x_+^{\pm n}$; $\beta = \delta x_-^{\pm n}$
then

$$L_{\pm n} = x_+^{1\mp n} \frac{\partial}{\partial x_+}, \quad \tilde{L}_{\pm n} = x_-^{1\pm n} \frac{\partial}{\partial x_-} \tag{34}$$

which are nothing but the generators of Virasoro algebra

$$[\tilde{L}_n,\tilde{L}_m] = (m-n)\tilde{L}_{m+n} \; ; \quad [L_m,\tilde{L}_n] = 0 \; ; \; [L_n,L_m] = (m-n)L_{m+n}.$$

Thus it is possible to express the infinite number of symmetries of the Liouville problem in terms of Virasoro algebra, which is intimately associated with the conformal invariance of the system, which is to be discussed in Sect. 6.

5. Riemann Hilbert Transform and Hidden Symmetries

Infinite dimensional Lie algebra plays an important role in the generation of hidden symmetries of nonlinear integrable systems. Such symmetry transformations have been obtained in two, three and four dimensional integrable systems. There are quite a variety of approaches to the study of such transformations. Here we shall elaborate on two such methodologies.

Let us consider the equations motion of Heisenberg spin chain [12]

$$\partial_t S = \frac{1}{2i} \partial_x [S,\partial_x S]$$

where $S = \sum_{i=1}^{3} S_i(x,t)\,\sigma_i$, $S^2 = 1$.

The Lax pair is given by

$$\partial_x Y^{(\lambda)} = -i\lambda\, S Y^{(\lambda)}$$

$$\partial_t Y^{(\lambda)} = (-\tfrac{\lambda}{2}[S,\partial_x S] + 2i\lambda^2 S)Y^{(\lambda)} . \tag{35}$$

Let us consider all such solutions $Y(\lambda)$ of the linear problem with the condition $Y^0(t,x) = 1$, and make the transformation

$$Y^{(\lambda)} \rightarrow (1 + \delta M_x)Y^{(\lambda)} ,$$

where δ is a small parameter and M_x is defined as

$$M_x Y^{(\lambda)} = -\frac{1}{2\pi i} \int_{C_0,\lambda} \frac{d\mu}{\mu}\cdot \frac{\lambda}{\lambda-\mu} \cdot V_x(\mu) Y^{(\lambda)}$$

$$V_x(\mu) = Y^{(\mu)} X(\mu) Y^{(\mu)-1} .$$

Coefficients of the Lax equations (35) transform as

$$M_x S = -\frac{1}{2\pi i}\int_{C_0} \frac{d\mu}{\mu}[V_x(\mu), S]$$

$$M_x[S,\partial_x S] = -\frac{1}{2\pi i}\int_{C_0}\Big[V_x(\mu), \quad [S, \partial_x S] - 4i\mu S\Big] .$$

Here C_0 denotes a circular contour in the complex μ plane with $\mu = 0$ as centre. It can be easily verified that $(1+\delta M_x)Y^{(\lambda)}$ solves the Lax equation upto δ^2 terms. So these are nothing but infinitesimal symmetry transformations of the Lax pair. To study the algebra generated by them we compute

$$[M_x, M_y] \tag{36}$$

for which we observe that

$$M_x(V_y(\mu)) = -\frac{1}{2\pi i}\int_{C_{0,\mu}} \frac{dp}{p}\,\frac{\mu}{\mu-p}[V_x(p), V_y(\mu)]$$

and hence we can write

$$\left([M_x,M_y]Y^{(\lambda)}\right)\left(Y^{(\lambda)}\right)^{-1}$$

$$= \frac{1}{2\pi i}\int_{C_{0,\lambda}} \frac{d\mu}{\mu}\,\frac{1}{2\pi i}\int_{C_{0;\lambda;\mu}} dp\,\Big\{\frac{\lambda}{\lambda-\mu}\,\frac{\mu}{p(\mu-p)}[V_x(p),V_y(\mu)]$$

$$- \frac{\lambda}{(\lambda-p)(\mu-p)}[V_y(p), V_x(\mu)]\Big\} .$$

Now, the contour $C_{0,\lambda,\mu}$ can be deformed into a contour C_0' lying inside $C_{0;\lambda}$, plus a contribution from $p = \mu$. In the second term $C_{0,\lambda}$ can be deformed in C_0' whence we get

$$\left([M_x,M_y]Y^{(\lambda)}\right)\left(Y^{(\lambda)}\right)^{-1} = -\frac{1}{2\pi i}\int_{C_{0,\lambda}} \frac{d\mu}{\mu}\,\frac{\lambda}{\lambda-\mu}[V_x(\mu),V_y(\mu)]$$

$$= +\frac{1}{2\pi i}\int_{C_{0;\lambda}} d\mu\,\frac{1}{2\pi i}\int_{C_0'} dp\,\frac{\lambda}{p(\lambda-\mu)(\mu-p)}[V_x(p),V_y(\mu)]$$

$$-\frac{1}{2\pi i}\int_{C_0'} d\lambda\,\frac{1}{2\pi i}\int_{C_{0,\lambda}} dp\,\frac{\lambda}{\mu(\lambda-p)(\mu-p)}[V_y(p),V_x(\mu)] \tag{37}$$

So we get

$$[M_x, M_y]Y^{(\lambda)} = -\frac{1}{2\pi i}\int \frac{d\mu}{\mu}\ \frac{\lambda}{\lambda-\mu}\ V^{\mu}_{[X,Y]}Y^{(\lambda)}$$

$$= M_{X,Y}\ Y^{(\lambda)}\ .$$

Thus they form a Lie algebra. Now, for the basis of the form

$$X_a^{(n)}(\lambda) = \lambda^{-m}\ \frac{\sigma_a}{2i}\ ; \qquad M_0^n = M_{X_a^n}$$

we get

$$[M_a^n, M_b^m] = \varepsilon_{abc}\ M_c^{n+m}\ .$$

This is actually the elegant formalism of Eichenherr [13] for the derivation of Kac-Moody algebra. Similar approach was also followed by Ueno and Nakamura [14]. But an important ingredient is lacking in this formalism. The formalism cannot give explicit grading of the generators of the infinite dimensional Lie algebra. So, at this point we can refer to a second method, where the gradation of Kac-Moody algebra is fixed by the imposition of the reduction condition associated with a particular nonlinear problem. We will now review the method in short.

Let us consider the Lax pair

$$\partial_x\psi = [u_0 + \lambda u_1]\psi = u\psi$$

$$\partial_t\psi = \lambda^{-1}V_1\psi = V\psi \quad . \tag{38}$$

It is easily seen that sine-Gordon equation is the result of Forbenius consistency of (38) if we also impose the reduction condition [15]

$$\begin{aligned} &\psi \in SL(2, C) \\ &(\psi(\lambda))^* = \psi(-\bar{\lambda})^{-1} \\ &\sigma_3\psi(\lambda)\sigma_3^{-1} = \psi(-\lambda)\ . \end{aligned} \tag{39}$$

The Riemann-Hilbert transformation is given by

$$\chi(\lambda) = 1 - \frac{1}{2\pi i}\int_C \frac{dt}{t-\lambda}\ \chi(t)[S(t)S(\lambda)^{-1} - 1]\ . \tag{40}$$

Let χ_+, χ_- denote the analytic parts of the transformation inside and outside C. We have introduced S as

$$\chi_+(\lambda) = \chi_-(\lambda) S$$

$$S = \psi(\lambda)\ \gamma(\lambda)\ \psi(\lambda)^{-1} .$$

Then the Riemann-Hilbert transformation keeps the Lax pair invariant if $\gamma(\lambda)$ obeys (39). We now pass over to the algebra from the group element by choosing infinitesimal $\gamma(\lambda)$ as [16]

$$\gamma(\lambda) = e^{i\Theta(\lambda)} \approx 1 + \Theta(\lambda)$$

$$\Theta(\lambda) = \sum_{-\infty}^{\infty} \lambda^n \Theta_n$$

whence due to (39) we must have

$$\mathrm{Tr}[\Theta(\lambda)] = 0$$

$$(\Theta(\lambda))^* = -\Theta(-\bar{\lambda})$$

$$\Theta_n^* = (-1)^{n+1}\ \Theta_n\ ;\ \sigma_3\Theta(\lambda)\ \sigma_3 = \Theta(-\lambda)$$

$$\sigma_3\Theta_n\sigma_3 = (-1)^n\ \Theta_n .$$

So a grading is fixed in powers of λ .
Hence we can write

$$\Theta(\lambda) = \Sigma\ C_k\ i\sigma_3\ \lambda^{2k} + \Sigma C'_k\ \sigma_1\ \lambda^{2k+1} + \Sigma\ C''_k\ \sigma_2\ \lambda^{2k+1} .$$

It then follows that these symmetries generate a subalgebra given by

$$[T_a^m , T_b^m] = \epsilon_{abc}\ T_e^{m+n}$$

$$T_e^m = \begin{cases} \lambda_i^m\ \sigma_3\delta_{a,3} & m \text{ even} \\ \lambda^m(\sigma_1\delta_{a,1} + \sigma_2\delta_{a,2}) & m \text{ odd} \end{cases} \qquad (41)$$

giving rise to the specific gradings of the generators of the algebra.

6. Bi-Hamiltonian Structure and Re-parameterization Invariance

Lastly, we discuss the role of such infinite dimensional Lie algebra in the study of the conformal invariance or reparameterization invariance on a circle S'. Already in Sect. 4, we have mentioned that the infinite dimensional algebra is related to conformal invariance in two dimensions. At this point it can be noted that the bi-Hamiltonian structure of integrable system is also related to such conformal invariance of the system [17].

To introduce the ideas involved, let us consider the KdV equation

$$u_t = u''' + 6uu' \quad (\ ' = \partial_x) \tag{42}$$

which is known to have the Hamiltonian structure

$$u_t = \mathbb{D} \frac{\delta H}{\delta u} .$$

Two possibilities arise for $\mathbb{D}$

$$\mathbb{D}^1 = \partial_x \ ; \quad H^1 = \frac{1}{2}\int \{2u^3 - (u')^2\}\, dx$$

and

$$\mathbb{D}^2 = \frac{1}{2}(\partial_x^3 + 4u\partial_x + 2u') \ ; \quad H^2 = \int u^2\, dx . \tag{43}$$

These give rise to the two different definitions of the Poisson bracket of two functionals

$$\{F,G\}_i = \int \frac{\delta F}{\delta u}\, \mathbb{D}^i\, \frac{\delta G}{\delta u}\, dx . \tag{44}$$

The KdV equation makes contact with the Virasoro algebra through the second Hamiltonian structure [18]. One supposes that the field is a periodic function of x; $x \in [0,2\pi]$ and expands it in Fourier series. That is, we set

$$u(x,t) = -\frac{12}{c}\sum_{-\infty}^{\infty} \ln e^{in\,x} + \frac{1}{2} \tag{45}$$

whence from

$$\left\{u(x),\ u(x')\right\}_2 = \frac{1}{2}(\partial_x^3 + 4u\,\partial_x + 2u')\partial(x - x') . \tag{46}$$

We get the following commutation rule for the generators L_n:

$$\{L_n, L_m\} = (n - m)L_{n+m} + \frac{c}{12}(n^3 - n)\delta_{n+m,0} . \tag{47}$$

An interesting outcome of such an observation is that one can develop a generic procedure to construct classical conformal algebra starting from multi-Hamiltonian system. Actually, the above story is a smaller part of a general framework formulated by Gelfand-Dikii. They showed that a class of coupled nonlinear evolution equations is generated if one starts from an arbitrary L-operator

$$L_n = \partial_x^n + \sum_{i=0}^{n-2} u_i \partial^i x \tag{48}$$

and uses the corresponding Lax equation

$$\partial_t L_n = \left[L_{n_+}^{k/n}, L_n\right] \tag{49}$$

where $L^{k/n}$ denotes the fractional power of the operator L introduced by Gelfand and Dikii. Furthermore, one can think of u to be a periodic function of x that is defined on a unit circle, then it has been observed that the operator L_2 transforms covariantly under the co-adjoint action of Diff.(S') and acts on densities of degree $-\frac{1}{2}$ [19]. Mathematically it means that under a reparametrization of the circle $x \to z(x)$

$$L_{(2)} \to (z')^{-3/2} L_2 (z')^{-\frac{1}{2}}$$

$$L_2 = \partial_x^2 + \bar{u} \tag{50}$$

$$\bar{u} = (z')^2 u(z(x)) + \frac{1}{2} S_2$$

$$S_2 = \frac{z'''}{z'} - \frac{3}{2}\left(\frac{z''}{z'}\right)^2$$

the Schwarzian derivative indicating that $\tilde{u}$ transforms as a density of degree 2 just like the energy-momentum tensor of a two-dimensional field theory denoted by T. On the other hand, it has been observed that the conformal invariance demands

$$T(z)T(\omega) \sim \frac{C/2}{(z-\omega)^4} + \frac{2T(\omega)}{(z-\omega)^2} + \frac{\partial_\omega T}{z-\omega} \tag{51}$$

which is nothing but the Poisson-bracket generated by second Hamiltonian structures. It has been seen that this analogy can be further extended to many more system of coupled nonlinear systems, and higher order non-linear equations. In a recent communication we have found a new tri-Hamiltonian system, where we have shown that a new extension of Virasoro algebra is possible.

7. Conclusion

In the above we have discussed some applications of infinite dimensional Lie algebra by demonstrating the various usages. Of course, it is only a tip of a big iceberg which is yet to be explored. But I think this will serve to induce other people and research workers to join the game.

Acknowledgment

It is a pleasure to thank Prof. M. Lakshmanan for inviting me to deliver this lecture.

References

[1] R. K. Bullough and P. J. Caudrey, eds., **Solitons**, Current Topics in Physics (Springer-Verlag, Berlin, New York, 1981); M. Lakshmanan, ed., **Solitons: Introduction and Applications** (Springer-Verlag, Heidelberg, 1987).
[2] M. J. Ablowitz, A. Ramani, H. Segur, Lett. Nuovo Cimento **23** (1978) 333; J. Math. Phys. **21** (1980) 715, 1006; J. Weiss, J. Math. Phys. **24** (1983) 1405; 25 (1984) 13; A. Roy Chowdhury and P. K. Chanda, J. Math. Phys. **27** (1987) 707; P. K. Chanda and A. Roy Chowdhury, J. Math. Phys. **29** (1988) 843; R. Sahadevan and M. Lakshmanan, Phys. Rev. **A31** (1985) 861; Phys. Lett. **101A** (1984) 189.
[3] G. L. Lamb, Jr., **Elements of Soliton Theory** (John Wiley, New York, 1980).
[4] D. Sattinger, Stud. Appl Math. **72** (1983) 65; S. Olefsson, J. Phys. **A22** (1989) 157; A. Roy Chowdhury and S. Roy, J. Math. Phys. **27** (1986) 2464.
[5] P. Goddard and E. Olive, Int. J. Mod. Phys. **A1** (1986) 303; V. Kac, **Infinite Dimensional Lie Algebra** (Birkhauser, Boston, 1983).
[6] H. D. Wahlquist and F. B. Estabrook, J. Math. Phys. **16** (1975) 1.
[7] C. Hoenselares, Prog. Theo. Phys. **74** (1985) 645.
[8] F. B. Estabrook and H. D. Wahlquist, J. Math. Phys. **17** (1976) 1293.
[9] H. N. Van Eck, Proc. Netherland Acad. Sci. **A86** (1983) 149.
[10] S. Roy and A. Roy Chowdhury, Int. J. Theo. Phys. **28** (1989) 845.
[11] G. Bhattacharya and H. Bohr, ICTP Preprint 81/82 (13).
[12] L. D. Faddeev and L. A. Takhtajan, **Hamiltonian Methods in the Theory of Soliton** (Springer-Verlag, Berlin, 1988).
[13] H. Eichenherr, CERN Preprint, TH 3491.
[14] K. Ueno and Y. Nakamura, Phys. Lett. **109B** (1982) 273.
[15] A. Mikailov, Physica **3D** (1981) 73.
[16] H. Bohr and A. Roy Chowdhury, J. Phys. **A18** (1985) 3423.
[17] J. L. Gervais, Phys. Lett. **160B** (1985) 277; J. L. Gervais and A. Neveu, Nucl. Phys. **B209** (1982) 125.
[18] C. A. Laberge and P. Matheiu, Phys. Lett. **B215** (1988) 718; P. Matheiu , J. Math. Phys. **29** (1988) 2499.
[19] T. Bakas, Nucl. Phys. **B302** (1988) 189; E. Witten, Comm. Math. Phys. **114** (1988) 1.

Generalised Lie Symmetries and Integrability of Coupled Nonlinear Oscillators with Two Degrees of Freedom

M. Senthilvelan and M. Lakshmanan

Department of Physics, Bharathidasan University,
Tiruchirapalli 620024, India

We consider the integrability property of coupled polynomial nonlinear oscillators from the point of view of symmetries, with two degrees of freedom as a prototype example. It is shown how invariance of the equations of motion under generalised Lie transformations can lead to the identification of integrable cases. The perturbed Kepler problem is treated explicitly. Results on other two-dimensional systems are presented briefly.

1. Introduction

In recent years, there has been a great deal of interest in finding the interrelation between the dynamical symmetries of the equations of motion and the integrability of the equivalent Lagrangian systems, at least for finite degrees of freedom, through invariance analysis [1]. The conventional invariance analysis of dynamical systems is usually restricted to point transformations alone [2,3]. However, the application of point symmetries does not in general play a vital role in nonlinear dynamical systems because it gives rise only to the time-translation symmetry in most cases (though there are exceptions) and thereby leading to energy conservation alone. Thus it will be advantageous to include velocity dependent terms also in the transformations in order to obtain more general symmetries. Such transformations are being studied vigorously by various authors in recent years in different contexts [4,5]. Recently the existence of such nontrivial generalised dynamical symmetries have been related to the integrability properties of the underlying Lagrangian systems [6]. Subsequently many Lagrangian systems have been identified to be integrable through invariance analysis [7]. In this paper we wish to give a brief review of the relation between the integrability and the dynamical symmetries by considering some additional two dimensional Lagrangian dynamical systems.

The organization of the paper is as follows. In Sect. 2 we briefly describe the invariance analysis applicable to the two dimensional

Symmetries and Singularity Structures

systems and the use of Noether's theorem for constructing the integrals of motion from the infinitesimal symmetries. In Sect. 3 the method is applied to the perturbed Kepler problem as an example and the parametric choices at which the nontrivial dynamical symmetries exist are isolated. The associated integrals of motion for each set of dynamical symmetries are constructed via Noether's theorem. Brief details of other integrable systems are presented in Table 1. Finally Sect. 4 contains a brief discussion of the results.

2. Invariance Analysis for Two Degrees of Freedom Systems [4,6]

2.1 *The Invariance Condition*

Let us consider a Lagrangian system with two degrees of freedom,

$$L = \frac{1}{2}(\dot{x}^2 + \dot{y}^2) - V(x,y) , \tag{2.1}$$

where V is the potential. The Euler-Lagrange's equations of motion are

$$\ddot{x} = \frac{\partial L}{\partial x} = \alpha_1 , \tag{2.2a}$$

$$\ddot{y} = \frac{\partial L}{\partial y} = \alpha_2 . \tag{2.2b}$$

Consider a one-parameter (ε) group of transformations from the variable x, y, t to the X,Y,T such that the new variables also depend on the time derivatives $\dot{x},\dot{y}$:

$$T = T(t,x,y,\dot{x},\dot{y},\varepsilon), \quad X = X(t,x,y,\dot{x},\dot{y},\varepsilon), \quad Y = Y(t,x,y,\dot{x},\dot{y},\varepsilon). \tag{2.3}$$

Then the time derivatives $\dot{X}$, $\dot{Y}$ take the following forms,

$$\dot{X} = \frac{dX}{dT} = \frac{\frac{\partial X}{\partial t} + \dot{x}\frac{\partial X}{\partial x} + \dot{y}\frac{\partial X}{\partial y} + \alpha_1\frac{\partial X}{\partial \dot{x}} + \alpha_2\frac{\partial X}{\partial \dot{y}}}{\frac{\partial T}{\partial t} + \dot{x}\frac{\partial T}{\partial x} + \dot{y}\frac{\partial T}{\partial y} + \alpha_1\frac{\partial T}{\partial \dot{x}} + \alpha_2\frac{\partial T}{\partial \dot{y}}} , \tag{2.4a}$$

$$\dot{Y} = \frac{dY}{dt} = \frac{\frac{\partial Y}{\partial t} + \dot{x}\frac{\partial Y}{\partial x} + \dot{y}\frac{\partial Y}{\partial y} + \alpha_1\frac{\partial Y}{\partial \dot{x}} + \alpha_2\frac{\partial Y}{\partial \dot{y}}}{\frac{\partial T}{\partial t} + \dot{x}\frac{\partial T}{\partial x} + \dot{y}\frac{\partial T}{\partial y} + \alpha_1\frac{\partial T}{\partial \dot{x}} + \alpha_2\frac{\partial T}{\partial \dot{y}}} . \tag{2.4b}$$

Equations (2.3) and (2.4) constitute a transformation from the space of variables $(t,x,y,\dot{x},\dot{y})$ to the space of variables $(T,X,Y,\dot{X},\dot{Y})$ and have the special property of mapping solution curves of (2.2) into curves

$$X = X(T) , \quad Y = Y(T) \tag{2.5}$$

in the $(T,X,Y,\dot{X},\dot{Y})$ space. In other words (2.5) must be a solution of the equations,

$$\ddot{X} = \alpha_1(X,Y) \ , \tag{2.6a}$$

$$\ddot{Y} = \alpha_2(X,Y) \ . \tag{2.6b}$$

Expanding now (2.3) about the identity $\varepsilon = 0$ we obtain the following infinitesimal transformations,

$$T = t + \varepsilon\xi(t,x,y,\dot{x},\dot{y}) + O(\varepsilon^2) \ ,$$

$$X = x + \varepsilon\eta_1(t,x,y,\dot{x},\dot{y}) + O(\varepsilon^2) \ ,$$

$$Y = y + \varepsilon\eta_2(t,x,y,\dot{x},\dot{y}) + O(\varepsilon^2) \ , \quad \varepsilon \ll 1 \tag{2.7}$$

where η_1, η_2 and ξ are the infinitesimals of the transformations of the variables x,y,t respectively. Substituting (2.7) in (2.4) and retaining the derivatives to first order in ε, we obtain

$$\dot{X} = \dot{x} + \varepsilon(\dot{\eta}_1 - \dot{\xi}\dot{x}) \ ,$$

$$\dot{Y} = \dot{y} + \varepsilon(\dot{\eta}_2 - \dot{\xi}\dot{y}) \ , \tag{2.8}$$

where

$$\eta_i = [\frac{\partial}{\partial t} + \dot{x}\frac{\partial}{\partial x} + \dot{y}\frac{\partial}{\partial y} + \alpha_1\frac{\partial}{\partial \dot{x}} + \alpha_2\frac{\partial}{\partial \dot{y}}]\eta_i \ , \ i = 1,2. \tag{2.9}$$

From (2.7) and (2.8) the infinitesimal generator of the group is,

$$E = \xi\frac{\partial}{\partial t} + \eta_1\frac{\partial}{\partial x} + \eta_2\frac{\partial}{\partial y} + (\dot{\eta}_1 - \dot{\xi}\dot{x})\frac{\partial}{\partial \dot{x}} + (\dot{\eta}_2 - \dot{\xi}\dot{y})\frac{\partial}{\partial \dot{y}} \ . \tag{2.10}$$

Using (2.7)-(2.9) in (2.2) one can check that the equations of motion (2.2) is invariant under the transformations (2.7) provided the following invariance conditions hold:

$$\ddot{\eta}_1 - \dot{x}\ddot{\xi} - 2\dot{\xi}\alpha_1 = E\{\alpha_1\} \ ,$$

$$\ddot{\eta}_2 - \dot{x}\ddot{\xi} - 2\dot{\xi}\alpha_2 = E\{\alpha_2\} \ . \tag{2.11}$$

2.2 Polynomial Forms of Transformations

The invariance conditions (2.11) form an incomplete system in ξ, η_1 and η_2. In order to solve (2.11) one has to consider specific forms for ξ, η_1 and η_2. One such choice is to assume ξ, η_1 and η_2 as polynomials in the velocities with undetermined coefficients and then determine the coefficients consistently. That is,

$$\xi = \sum_{i,j=0}^{N} a_{ij}\dot{x}^i\dot{y}^j \ , \quad \eta_1 = \sum_{i,j=0}^{N} b_{ij}\dot{x}^i\dot{y}^j \ , \quad \eta_2 = \sum_{i,j=0}^{N} c_{ij}\dot{x}^i\dot{y}^j . \tag{2.12}$$

where the coefficients a_{ij}, b_{ij} and c_{ij} are functions of x,y and t only.

2.3 Linear Forms

First let us consider the simple case in which the infinitesimals are linear in the velocities,

$$\xi = a_1 + a_2\dot{x} + a_3\dot{y} \ ,$$

$$\eta = b_1 + b_2\dot{x} + b_3\dot{y} \ ,$$

$$\eta = c_1 + c_2\dot{x} + c_3\dot{y} \ , \tag{2.13}$$

where a_i, b_i, c_i , i = 1,2,3 are functions of x,y,t only. Substituting (2.13) into (2.11) and equating the various powers of $\dot{x}^m\dot{y}^n$ to zero, we obtain a system of 24 partial differential equations, which may be solved consistently to determine the unknown coefficients a_i, b_i, c_i, i = 1,2,3. However from the practical point of view, we do find that it is sufficient to assume that (i) $\xi = 0$, η_1, η_2: independent of t and (ii) ξ = constant, $\eta_1 = \eta_2 = 0$.

In the case of $\xi = 0$, η_1, η_2 independent of t, the consistency conditions become,

$$a_1 = a_2 = a_3 = 0 \ , \tag{2.14}$$

$$b_{1xx} = 0 \ , \quad b_{1xy} = 0 \ , \quad b_{1yy} = 0 \ ,$$

$$b_{2xx} = 0 \ , \quad b_{3yy} = 0 \ , \quad b_{2yy} + 2b_{3xy} = 0, \quad b_{3xx} + 2b_{2xy} = 0. \tag{2.15}$$

$$c_{1xx} = 0, \quad c_{1xy} = 0, \quad c_{1yy} = 0 \ ,$$

$$c_{2xx} = 0, \quad c_{3yy} = 0, \quad c_{2yy} + 2c_{3xy} = 0, \quad c_{3xx} + 2c_{2xy} = 0. \tag{2.16}$$

$$\alpha_1 b_{1x} + \alpha_2 b_{1y} - \alpha_{1x} b_1 - \alpha_{1y} c_1 = 0 \ ,$$

$$3\alpha_1 b_{2x} + \alpha_{2x} b_3 + 2\alpha_2 b_{3x} + \alpha_2 b_{2y} - \alpha_{1y} c_2 = 0 \ ,$$

$$\alpha_{1y}b_2 + 2\alpha_1 b_{2y} + \alpha_{2y}b_3 + 3\alpha_2 b_{3y} + \alpha_1 b_{3x} - \alpha_{1x}b_3 - \alpha_{1y}c_3 = 0. \quad (2.17a)$$

$$\alpha_1 c_{1x} + \alpha_2 c_{1y} - \alpha_{2x}b_1 - \alpha_{2y}c_1 = 0 ,$$

$$\alpha_{1x}c_2 + 3\alpha_1 c_{2x} + \alpha_{2x}c_3 + 2\alpha_2 c_{3x} + \alpha_2 c_{2y} - \alpha_{2x}b_2 - \alpha_{2y}c_2 = 0,$$

$$\alpha_{1y}c_2 + 2\alpha_1 c_{2y} + \alpha_1 c_{3x} + 3\alpha_2 c_{3y} - \alpha_{2x}b_3 = 0 . \quad (2.17b)$$

Solving (2.14) we get,

$$b_1 = b_{10}x + b_{11}y + b_{12} ,$$

$$b_2 = b_{20}y^2 + b_{21}xy + b_{22}x + b_{23}y + b_{24} ,$$

$$b_3 = -b_{21}x^2 - b_{20}xy + b_{31}x + b_{32}y + b_{33} , \quad (2.18a)$$

$$c_1 = c_{10}x + c_{11}y + c_{12} ,$$

$$c_2 = c_{20}y^2 + c_{21}xy + c_{22}x + c_{23}y + c_{24} ,$$

$$c_3 = -c_{21}x^2 - c_{20}xy + c_{31}x + c_{32}y + c_{33} , \quad (2.18b)$$

where b_{ij}'s and c_{ij}'s are constants. Substituting the forms (2.18) in (2.17) with a given potential V, we get six algebraic equations in x and y. Thus the problem has been reduced to merely solving six algebraic equations out of the 24 partial differential equations of the time dependent case.

If we equate the various powers of $x^m y^n$ to zero in the six algebraic equations we may find the parametric choices of the potential for which nontrivial dynamical symmetries exist [6].

2.4 Cubic Forms

In the case where ξ, η_1 and η_2 are cubic polynomials in $\dot{x},\dot{y}$, we have

$$\xi = a_1 + a_2\dot{x}+a_3\dot{y}+a_4\dot{x}^2+a_5\dot{x}\dot{y}+a_6\dot{y}^2+a_7\dot{x}^3+a_8\dot{x}^2\dot{y} + a_9\dot{x}\dot{y}^2+a_{10}\dot{y}^3,$$

$$\eta_1 = b_1+b_2\dot{x}+b_3\dot{y}+b_4\dot{x}^2+b_5\dot{x}\dot{y}+b_6\dot{y}^2+b_7\dot{x}^3+b_8\dot{x}^2\dot{y}+b_9\dot{x}\dot{y}^2+b_{10}\dot{y}^3 ,$$

$$\eta_2 = c_1+c_2\dot{x}+c_3\dot{y}+c_4\dot{x}^2+c_5\dot{x}\dot{y}+c_6\dot{y}^2+c_7\dot{x}^3+c_8\dot{x}^2\dot{y}+c_9\dot{x}\dot{y}^2+c_{10}\dot{y}^3, \quad (2.19)$$

where a_{ij}, b_{ij} and c_{ij}'s are functions of (t,x,y) only. If we substitute the above form in the invariance condition (2.11) and repeat the above analysis we get an overdetermined system of 48 partial differential equations. But again if we can consider the case $\xi = 0$, η_1 and η_2 time independent , then we get the following consistency conditions:

$$a_1 = a_2 = a_3 = a_4 = a_5 = a_6 = a_7 = a_8 = a_9 = a_{10} = 0. \qquad (2.20a)$$

$$b_{4xx} = 0, \quad b_{6yy} = 0, \quad b_{7xx} = 0, \quad b_{10yy} = 0 ,$$

$$b_{5xx}+2b_{4xy} = 0, \quad b_{5yy}+2b_{6xy} = 0, \quad b_{8xx}+2b_{7xy} = 0,$$

$$b_{9yy}+2b_{10xy} = 0, \quad b_{4yy}+b_{6xx}+2b_{5xy} = 0, \quad b_{7yy}+b_{9xx}+2b_{8xy} = 0,$$

$$b_{8yy}+b_{10xx}+2b_{9xy} = 0 . \qquad (2.20b)$$

$$c_{4xx} = 0, \quad c_{6yy} = 0, \quad c_{7xx} = 0, \quad c_{10yy} = 0 ,$$

$$c_{5xx}+2c_{4xy}= 0, \quad c_{5yy}+2c_{6xy}= 0, \quad c_{8xx}+2c_{7xy}= 0,$$

$$c_{9yy}+2c_{10xy} = 0 , \quad c_{4yy}+c_{6xx}+2c_{5xy} = 0, \quad c_{7yy}+c_{9xx}+2c_{8xy}= 0,$$

$$c_{8yy}+c_{10xx}+2c_{9xy} = 0 . \qquad (2.20c)$$

$$b_{1xx}+5\alpha_1 b_{4x}+\alpha_{1x}b_4+\alpha_2 b_{4y}+2\alpha_2 b_{5x}+ \alpha_{2x}b_5-\alpha_{1y}c_4 = 0 ,$$

$$b_{1yy}+2\alpha_1 b_{5y}+\alpha_{1y}b_5+2\alpha_{2y}b_6+5\alpha_2 b_{6y}+\alpha_1 b_{6x}-\alpha_{1x}b_6-\alpha_{1y}c_6 = 0 ,$$

$$2b_{1xy}+4\alpha_1 b_{4y}+2\alpha_{1y}b_4+3\alpha_1 b_{5x}+3\alpha_2 b_{5y}+\alpha_{2y}b_5+2\alpha_{2x}b_6+4\alpha_2 b_{6x}-\alpha_{1y}c_5=0,$$

$$b_{2xx}+2\alpha_{1x}b_7+7\alpha_1 b_{7x}+\alpha_2 b_{7y}+\alpha_{2x}b_8+2\alpha_2 b_{8x}-c_7\alpha_{1y} = 0 ,$$

$$b_{3yy}+2\alpha_1 b_{9y}+\alpha_{1y}b_9+3\alpha_{2y}b_{10}+7\alpha_2 b_{10y}+\alpha_1 b_{10x}-\alpha_{1x}b_{10}-\alpha_{1y}c_{10} = 0,$$

$$b_{3xx}+2b_{2xy}+3\alpha_{1y}b_7+6\alpha_1 b_{7y}+\alpha_{1x}b_8+5\alpha_1 b_{8x}+\alpha_{2y}b_8+3\alpha_2 b_{8y}+2\alpha_{2x}b_9 + 4\alpha_2 b_{9x}-\alpha_{1y}c_8 = 0 ,$$

$$b_{2yy}+2b_{3xy}+2\alpha_{1y}b_8+4\alpha_1 b_{8y}+3\alpha_1 b_{9x}+2\alpha_{2y}b_9+5\alpha_2 b_{9y}+3\alpha_{2x}b_{10} + 6\alpha_2 b_{10x}-\alpha_{1y}c_9 = 0 ,$$

$$2\alpha_1^2 b_4+2\alpha_1\alpha_2 b_5+2\alpha_2^2 b_6+\alpha_1 b_{1x}+\alpha_2 b_{1y}-\alpha_{1x} b_1-\alpha_{1y} c_1 = 0 ,$$

$$6\alpha_1^2 b_7+4\alpha_1\alpha_2 b_8+2\alpha_2^2 b_9+3\alpha_1 b_{2x}+\alpha_2 b_{2y}+\alpha_{2x} b_3+2\alpha_2 b_{3x}-\alpha_{1y} c_2 = 0,$$

$$6\alpha_2^2 b_{10}+4\alpha_1\alpha_2 b_9+\alpha_{1y} b_2+2\alpha_1 b_{2y}+\alpha_{2y} b_3+3\alpha_2 b_{3y}+2\alpha_1^2 b_8+\alpha_1 b_{3x}$$

$$- \alpha_{1x} b_3-\alpha_{1y} c_3 = 0 . \qquad (2.20d)$$

$$c_{1xx}+5\alpha_1 c_{4x}+2\alpha_{1x} c_4+\alpha_2 c_{4y}+2\alpha_2 c_{5x}+\alpha_{2x} c_5-\alpha_{2x} b_4-\alpha_{2y} c_4 = 0,$$

$$c_{1yy}+2\alpha_1 c_{5y}+\alpha_{1y} c_5+\alpha_{2y} c_6+5\alpha_2 c_{6y}+\alpha_1 c_{6x}-\alpha_{2x} b_6 = 0 ,$$

$$2c_{1xy}+4\alpha_1 c_{4y}+2\alpha_{1y} c_4+3\alpha_1 c_{5x}+\alpha_{1x} c_5+3\alpha_2 c_{5y}+2\alpha_{2x} c_6+4\alpha_2 c_{6x}-\alpha_{2x} b_5=0,$$

$$c_{2xx}+3\alpha_{1x} c_7+7\alpha_1 c_{7x}+\alpha_2 c_{7y}+\alpha_{2x} c_8+2\alpha_2 c_{8x}-\alpha_{2x} b_7-\alpha_{2y} c_7 = 0 ,$$

$$c_{3yy}+2\alpha_1 c_{9y}+\alpha_{1y} c_9+2\alpha_{2y} c_{10}+7\alpha_2 c_{10y}+\alpha_1 c_{10x}-\alpha_{2x} b_{10} = 0,$$

$$c_{3xx}+2c_{2xy}+3\alpha_{1y} c_7+6\alpha_1 c_{7y}+2\alpha_{1x} c_8+5\alpha_1 c_{8x}+3\alpha_2 c_{8y}+2\alpha_{2x} c_9$$

$$+ 4\alpha_2 c_{9x}-\alpha_{2x} b_8 = 0 ,$$

$$c_{2yy}+2c_{3xy}+2\alpha_{1y} c_8+4\alpha_1 c_{8y}+3\alpha_1 c_{9x}+\alpha_{1x} c_9+\alpha_{2y} c_9+5\alpha_2 c_{9y}$$

$$+ 3\alpha_{2x} c_{10}+6\alpha_2 c_{10x}-\alpha_{2x} b_9 = 0 ,$$

$$2\alpha_1^2 c_4+2\alpha_1\alpha_2 c_5+2\alpha_2^2 c_6+\alpha_1 c_{1x}+ \alpha_2 c_{1y}-\alpha_{2x} b_1-\alpha_{2y} c_1 = 0 ,$$

$$6\alpha_1^2 c_7+4\alpha_1\alpha_2 c_8+2\alpha_2^2 c_9+\alpha_{1x} c_2+3\alpha_1 c_{2x}+\alpha_2 c_{2y}+\alpha_{2x} c_3$$

$$+ 2\alpha_2 c_{3x}-\alpha_{2x} b_2-\alpha_{2y} c_2 = 0 ,$$

$$6\alpha_2^2 c_{10}+4\alpha_1\alpha_2 c_9+\alpha_{1y} c_2+2\alpha_1 c_{2y}+3\alpha_2 c_{3y}+\alpha_1 c_{3x}+2\alpha_1^2 c_8-\alpha_{2x} b_3 = 0. \qquad (2.20e)$$

Solving (2.20) we get

$$b_4 = b_{41} y^3+b_{42} xy^2+b_{43} y^2+b_{44} xy+b_{45} x+b_{46} y+b_{47} ,$$

$$b_5 = -2b_{42} x^2 y-2b_{41} xy^2-b_{44} x^2-b_{51} y^2+b_{52} xy+b_{53} x+b_{54} y+b_{55} ,$$

$$b_6 = b_{42} x^3+b_{41} x^2 y+b_{61} x^2+b_{51} xy+b_{62} x+b_{63} y+b_{64} ,$$

$$b_7 = b_{70}xy^3+b_{71}xy^2+b_{72}xy+b_{73}x-b_{100}y^4-b_{80}y^3+b_{75}y^2+b_{76}y+b_{77},$$

$$b_8 = -3b_{70}x^2y^2-2b_{71}x^2y-b_{72}x^2+3b_{100}xy^3+b_{80}xy^2+b_{81}xy$$

$$+ b_{82}x+b_{101}y^3+b_{83}y^2+b_{84}y+b_{85} ,$$

$$b_9 = -3b_{100}x^2y^2-2b_{101}xy^2-b_{102}y^2+3b_{70}x^3y+b_{80}x^2y+b_{91}xy+b_{92}y+b_{71}x^3$$

$$+ b_{93}x^2+b_{94}x+b_{95} ,$$

$$b_{10} = b_{100}x^3y+b_{101}x^2y+b_{102}xy+b_{103}y-b_{70}x^4-b_{80}x^3+b_{105}x^2+b_{106}x+b_{107}.$$

(2.21a)

$$c_4 = c_{41}y^3+c_{42}xy^2+c_{43}y^2+c_{44}xy+c_{45}x+c_{46}y+c_{47} ,$$

$$c_5 = -2c_{42}x^2y-2c_{41}xy^2-c_{44}x^2-c_{51}y^2+c_{52}xy+c_{53}x+c_{54}y+c_{55} ,$$

$$c_6 = c_{42}x^3+c_{41}x^2y+c_{61}x^2+c_{51}xy+c_{62}x+c_{63}y+c_{64} ,$$

$$c_7 = c_{70}xy^3+c_{71}xy^2+c_{72}xy+c_{73}x-c_{100}y^4-c_{80}y^3+c_{75}y^2+c_{76}y+c_{77} ,$$

$$c_8 = -3c_{70}x^2y^2-2c_{71}x^2y-c_{72}x^2+3c_{100}xy^3+c_{80}xy^2+c_{81}xy+c_{82}x+c_{101}y^3$$

$$+ c_{83}y^2+c_{84}y+c_{85} ,$$

$$c_9 = -3c_{100}x^2y^2-2c_{101}xy^2-c_{102}y^2+3c_{70}x^3y+c_{80}x^2y+c_{91}xy$$

$$+ c_{92}y+c_{71}x^3+c_{93}x^2+c_{94}x+c_{95} ,$$

$$c_{10} = c_{100}x^3y+c_{101}x^2y+c_{102}xy+c_{103}y-c_{70}x^4-c_{80}x^3$$

$$+ c_{105}x^2+c_{106}x+c_{107} , \qquad (2.21b)$$

where the various coefficients b_{ij}, c_{ij}, b_{ijk}, c_{ijk} are constants. As before using the forms (2.21) in (2.20d,e), we can find the form of the potentials admitting nontrivial dynamical symmetries [6].

Obviously this procedure can be extended to any higher degree forms of ξ, η_i, i = 1,2.

2.5 *Integrals of Motion*

Having obtained the explicit forms of infinitesimal symmetries, one can proceed to find the associated integrals of motion which are in involution, provided they exist. It is appropriate to mention here that not all the symmetries leave the Lagrangian invariant even though they leave the equation of motion invariant. For those sets of symmetries which leave the Lagrangian invariant (for non-Noetherian symmetries, we refer to [8]),

$$I = (\xi\dot{x} - \eta_1)\frac{\partial L}{\partial \dot{x}} + (\xi\dot{y} - \eta_2)\frac{\partial L}{\partial \dot{y}} - \dot{\xi}L + f , \tag{2.22}$$

where the function f(x,y) is to be determined from the equation

$$E\{L\} + \dot{\xi}L = \frac{\partial f}{\partial x}\dot{x} + \frac{\partial f}{\partial y}\dot{y} \tag{2.23}$$

so that I is an integral of motion. Obviously for the case, ξ = constant, $\eta_1, \eta_2 = 0$, the function f(x,y) turns out to be the potential V so that I = H, the Hamiltonian. It is important to note that the connection between the dynamical symmetries and the conserved quantities is not too direct: for instance, there always exists an infinite number of linearly independent dynamical symmetries which have no independent integrals of motion. Sarlet and Cantrijn have made a detailed study on the applicability of Noether's theorem to dynamical symmetries [5]. However, in all the systems below, we do find explicit involutive integrals of motion.

3. Perturbed Kepler Problem

In this section, we apply the above theory to find the integrable cases and the associated dynamical symmetries of the perturbed Kepler problem [9],

$$V = \frac{-g}{r} + ax^N + by^N , \tag{3.1}$$

where $r = (x^2+y^2)^{\frac{1}{2}}$ and a, b and g are constants.

Now substituting (3.1) with (2.18) in (2.17), we end up with six algebraic equations. Equating the various powers of $x^m y^n$ to zero in each equation and solving them finally we get the following four equations:

$$N(N-1)bc_{10}xy^{N-2} - Nac_{10}x^{N-1} = 0 ,$$

$$N(N-1)ac_{10}yx^{N-2} - Nbc_{10}y^{N-1} = 0 ,$$

$$3(b_{21}y - 2b_{31})Nax^{N-1} - 3(b_{21}x - 2b_{30})Nby^{N-1} - (b_{21}x - 2b_{30})$$

Table 1. Infinitesimal symmetries and integrals of motion for some dynamical systems

Case	Potential V	Parametric restrictions	Symmetries	Integrals of motion
1	$\frac{-g}{(x^2+y^2)^{\frac{1}{2}}} + ax^N+by^N$	(i) $N = 2,\ a = b$	$\xi = 0$ $\eta_1 = -y^2\dot{x}+xy\dot{y}$ $\eta_2 = xy\dot{x}-x^2\dot{y}$	$I_1 = (y\dot{x} - x\dot{y})$
		(ii) $N = 2,\ a = 4b$	$\xi = 0$ $\eta_1 = y\dot{y}$ $\eta_2 = y\dot{x}-2x\dot{y}$	$I_2 = [\dot{y}(y\dot{x}-x\dot{y})+\frac{gx}{(x^2+y^2)^{\frac{1}{2}}}+2bxy^2]$
		(iii) $N=1$, a,b arbitrary	$\xi = 0$ $\eta_1=-2y\dot{x}+(\frac{a}{b}y+x)\dot{y}$ $\eta_2=(x+\frac{a}{b}y)\dot{x}-\frac{2a}{b}x\dot{y}$	$I_3=(-b\dot{x}+a\dot{y})(y\dot{x}-x\dot{y})+\frac{g(ax+by)}{(x^2+y^2)^{\frac{1}{2}}}$ $+ (a^2+b^2)(x^2+y^2)$
2	$Ax^2+By^2+\alpha x^6+\beta y^6$ $+\delta_1 x^4y^2+\delta_2 x^2y^4$	(i) $A=B, \alpha=\beta, \delta_1=\delta_2=3\alpha$	$\xi = 0$ $\eta_1=2y(\dot{x}y-x\dot{y})$ $\eta_2=2x(x\dot{y}-y\dot{x})$	$I_1= (x\dot{y}-y\dot{x})^2$
		(ii) $A=B,\ \alpha=\beta, \delta_1=\delta_2=15\alpha$	$\xi = 0$ $\eta_1=k\dot{y}$ $\eta_2=k\dot{x}$	$I_2=\dot{x}\dot{y}+2Axy+6\alpha x^5y+6\alpha xy^5+20\alpha x^3y^3$

		(iii) $A=4B,\ \alpha=64\beta,\ \delta_1=80\beta,\ \delta_2=24\beta$	$\xi = 0$ $\eta_1=y\dot{y}$ $\eta_2=y\dot{x}-2x\dot{y}$	$I_3=(y\dot{x}-x\dot{y})\dot{y}+2(B+3\beta y^4+16\beta x^4 + 16\beta x^2y^2)xy^2$
3	$(x^2+y^2)^2+Ax^2+By^2 + \frac{C}{x^2} + \frac{D}{y^2}$	A,B,C,D arbitrary	$\xi = 0$ $\eta_1=[y^2+(B-A)]\dot{x}-xy\dot{y}$ $\eta_2=x(x\dot{y}-y\dot{x})$	$I=(y\dot{x}-x\dot{y})^2+2C(\frac{y}{x})^2+2D(\frac{x}{y})^2+(B-A)[\dot{x}^2+2x^4+2x^2y^2+2Ax^2+\frac{2C}{x^2}]$
4	$Ax^5+Bx^3y^2+Cxy^4$	(i) $B=10A,\ B=2C$	$\xi = 0$ $\eta_1=k\dot{y}$ $\eta_2=k\dot{x}$	$I_1=\dot{x}\dot{y}+(x^4y+2x^2y^3+y^5/5)$
		(ii) $A=B,\ B=\frac{16}{3}C$	$\xi = 0$ $\eta_1=y\dot{y}$ $\eta_2=y\dot{x}-2x\dot{y}$	$I_2=2y\dot{x}\dot{y}-2x\dot{y}^2+\frac{1}{3}(Cy^6+12Cx^2y^4+16Cx^4y^2)$
5	$Ax^2+By^2+Cx^4+Dx^2y^2 +Ey^4+\frac{F}{x^2}+\frac{G}{x^4}+\frac{Hy^2}{x^6}$	$B=4A,D=12C,H=4G,E=16C$	$\xi = 0$ $\eta_1=-2y\dot{x}+x\dot{y}$ $\eta_2=x\dot{x}$	$I=y\dot{x}^2-x\dot{x}\dot{y}+[\frac{2Fy}{x^2}+\frac{4Gy}{x^4}+\frac{2Hy^3}{x^6} - 2Ax^2y-4Cx^4y-\frac{2D}{3}x^2y^3]$
6	$Ax^2+By^2+Cx+\frac{D}{y^2}$	$B = 4A$	$\xi = 0$ $\eta_1=\dot{y}^2+k_1\dot{x}-2By^2+\frac{2D}{y^2}$ $\eta_2=k_2\dot{y}+2\dot{x}\dot{y}+8Bxy+Cy$	$I=\dot{x}\dot{y}^2+8Bxy\dot{y}+Cy\dot{y}-2By^2\dot{x}+\frac{2D}{y^2}\dot{x} +[k_1(\dot{x}^2+8Bx^2+2Cx) +k_2(\dot{y}^2+2By^2+\frac{2D}{y^2})]$

$$N(N-1)by^{N-1} + 2b_{31}N(N-1)bxy^{N-2} + (b_{21}y - 2b_{31})N(N-1)ax^{N-1}$$

$$- 2b_{30}N(N-1)ayx^{N-2} = 0 \; ,$$

$$3(c_{20}y + c_{22})Nax^{N-1} + (c_{20}y + c_{22})N(N-1)ax^{N-1} + c_{23}N(N-1)ayx^{N-2}$$

$$- 3c_{20}Nbxy^{N-1} - 3c_{23}Nby^{N-1} - (c_{20}x + c_{23})N(N-1)by^{N-1}$$

$$- c_{22}N(N-1)bxy^{N-2} = 0 \; , \qquad (3.2)$$

with

$$b_{21} = -2c_{20} \; , \quad b_{23} = -2b_{31} \; , \quad b_{31} = c_{22}, \; b_{24} = c_{32} \; , \; b_{23} = -2c_{22},$$

$$c_{31} = -2c_{23} \; , \quad c_{23} = b_{30} \; , \; c_{31} = -2b_{30} \; .$$

The above four equations are not in general compatible except for some particular values of N, a and b. Solving them we can isolate the following choices:

Case 1: $N = 2, \; a = b$,

Case 2: $N = 2, \; a = 4b$,

Case 3: $N = 1$, a, b arbitrary.

The corresponding dynamical symmetries are:

Case 1: $\eta_1 = -y^2\dot{x} + xy\dot{y}$, $\quad \eta_2 = xy\dot{x} - x^2\dot{y}$

Case 2: $\eta_1 = y\dot{y}$ $\quad \eta_2 = y\dot{x} - 2x\dot{y}$,

Case 3: $\eta_1 = -2y\dot{x} + \{(a/b)y + x\}\dot{y}$, $\eta_2 = \{x + (a/b)y\}\dot{x} - (2a/b)x\dot{y}$.

It is of interest to note that the above three parametric choices are exactly the same ones identified by Yoshida [9] through Ziglin's theorem approach. The associated integrals of motion using (2.22) are presented in Table 1.

4. Conclusion

In this paper we have pointed out how the method of invariance analysis involving generalized Lie symmetries enables one to identify the integrable dynamical systems involving two degrees of freedom. The procedure can be extended to three and higher dimensional systems as well without much difficulty. It will be of interest to see how nonpolynomial forms in velocities can be handled within this formalism.

References

[1] M. Lakshmanan, in **Nonlinear Physics**, ed. Yunbo Zeng et al. (Springer-Verlag, Heidelberg, 1990), in press.

[2] C. E. Wulfman and B. G. Wybourne, J. Phys. A: Math. Gen. **9** (1976) 507.

[3] M. Lutzky, J. Phys. A: Math. Gen. **11** (1978) 249.

[4] M. Lutzky, J. Phys. A: Math. Gen. **12** (1979) 973.

[5] W. Sarlet and F. Cantrijn, Siam Rev. **23** (1981) 487.

[6] R. Sahadevan and M. Lakshmanan, J. Phys. A: Math. Gen. **19** (1986) L949.

[7] M. Lakshmanan and R. Sahadevan, J. Math. Phys. (submitted).

[8] M. Lutzky, Phys. Lett. **72A** (1979) 86; **75A** (1979) 8; J. Phys. A: Math. Gen. **15** (1982) L87.

[9] H. Yoshida, Phys. Lett. **A120** (1987) 388.

Aspects of Symmetries of Dissipative Systems

G. Ambika, K. Babu Joseph, and K.M. Valsamma

Department of Physics, Cochin University of Science and Technology,
Cochin 682022, India

A dissipative system is characterised by a set of first order non-linear differential equations for the order parameters. We discuss translation, rotation and scale symmetries in order parameter space (OPS). A local criterion for spontaneous symmetry breaking (SSB) is introduced in terms of a certain exponent vanishing or becoming positive at the given point. It is shown that while translation and rotation symmetries may be spontaneously broken at particular points, dynamic scale symmetry cannot get broken. When the theory is applied to the Lorenz model, areas of SSB become identified in the OPS. This helps one to chart the progress of the system towards chaotic or non-chaotic behaviour.

1. Introduction

One of the well-known phenomena in dissipative systems is pattern formation which is normally associated with the presence of instabilities. Various physical as well as non-physical, including biological, systems belong to this class. Let us consider, for example, a liquid in a vessel to be heated from below. Beyond a critical temperature difference between the lower and upper surfaces, the initial passive state gets destabilised and is replaced by a roll pattern or hexagonal pattern discovered by Benard [1]. The actual pattern that is formed is determined by the vessel geometry. When the heating is continued further and further, the liquid may undergo a sequence of instabilities leading to oscillatory rolls and finally to chaos [2]. In the mathematical representation of a chaotic system the transition to chaos takes place through a number of bifurcations, each of which is sparked by a change in the control parameter. There is speculation in the literature about a possible relationship between bifurcations and spontaneous symmetry breaking (SSB). The factor responsible for symmetry breaking is the instabilities which cause bifurcations [3]. From this it is clear that symmetries and their breaking constitute an important problem in dissipative dynamics and in general, in synergetics [4].

Research Reports in Physics Symmetries and Singularity Structures
Editors: M. Lakshmanan · M. Daniel

We consider dissipative systems modeled by a first order autonomous system of differential equations:

$$\dot{x} = F_\lambda(x) , \quad (1)$$

where λ is the control parameter and, by analogy with phase transition theory, the column vector x will be called the order parameter vector. In this work we deal with the case where both x and F_λ belong to R^N.

A symmetry of the system (1) is a mapping

$$x \rightarrow x' , \quad F_\lambda(x) \rightarrow F'_\lambda(x') \quad (2)$$

that leaves (1) invariant:

$$\dot{x}' = F'_\lambda(x') \quad (3)$$

for a fixed value of λ. Traditionally the x-$\dot{x}$ space is called the phase-space of the system; we shall employ the term, **order parameter space** (OPS) for the x subspace of the phase space. In the present work we are mostly concerned with time dependent translations, rotations and scale transformations of the OPS.

If Σ denotes translation, rotation or a combination thereof, its effect on x is defined by

$$x \xrightarrow{\Sigma} x' , \quad F_\lambda(x) \xrightarrow{\Sigma} F_\lambda(x') . \quad (4)$$

Since Σ is continuously implementable an infinitesimal version can always be considered:

$$x \xrightarrow{\delta\Sigma} x + \delta x , \quad F_\lambda(x) \xrightarrow{\delta\Sigma} F_\lambda(x + \delta x) . \quad (5)$$

For fixed x an infinitesimal symmetry operation of this kind is a fluctuation. A fluctuation δx is said to be a normal mode fluctuation if it can be represented in the form

$$\delta x = \delta x_0 \exp(\omega_f t) , \quad (6)$$

where ω_f is called the fluctuation or normal mode exponent. As discussed in detail later the amplitude δx_0 possesses properties determined by the type of symmetry operation. If $\omega_f \geq 0$ the corresponding symmetry Σ is said to be spontaneously broken at the reference point $x = \xi$; if $\omega_f < 0$, Σ remains unbroken at ξ. It will be seen that $\omega_f = \omega_f(\xi,\lambda)$. Clearly the above criterion is only of local significance. But when the stability of a specified kind of symmetry is examined in a finite region, certain regularities may be revealed in the distribution of the particular form of stability.

The phrase, **spontaneous breaking,** is herein used in the sense that, even though the original system of equations remains invariant, no stable trajectory may be available through the point in question. In standard stability analysis fluctuations are considered about a critical or fixed point x_c defined by $\dot{x}_c = 0$. A critical point will be asymptotically stable or unstable depending on the value of the control parameter. Our motivation is not an investigation of the stability of critical points, which of course will form a special case of the generalised stability approach developed in this paper.

To sharpen the usage of the paradigm of SSB in dissipative dynamics some more jargon needs to be introduced. The point $x = 0$ is called the ordinary equilibrium or ordinary vacuum. Any other critical point $x_c = \xi$, $\xi = 0$, is called a higher equilibrium or higher vacuum. It may be verified by means of an example such as the Lorenz model (to be treated in section 4) that fluctuations around a higher equilibrium may not possess all the symmetries exhibited by those around the ordinary equilibrium. In the Lorenz model the order parameter x is a three vector (x_1,x_2,x_3) and the Lorenz equations possess the inversion symmetry $(x_1,x_2) \rightarrow (-x_1,-x_2)$, $x_3 \rightarrow x_3$, besides translation and rotation invariances. However, this discrete symmetry is maintained only by fluctuations about the ordinary equilibrium (0,0,0) but is broken in other cases. This is an almost trivial example for SSB. We argue that not only are discrete symmetries broken spontaneously around higher equilibria but continuous symmetries such as translation, rotation and scaling may get broken in certain regions of OPS. Such regions may be identified by means of a linear stability analysis appropriate to the symmetry in question. The unravelling of such regions in specific models would strengthen the view that chaos is not a catastrophe but is a slowly developed state of SSB.

The present paper consists of five sections. In Section 2 we introduce translational and rotational symmetries in OPS and Section 3 discusses scale invariance. In Section 4 the Lorenz model is treated as an illustration from the point of view of translations and rotations. The conclusions are summed up in Section 5.

2. Translational and Rotational Symmetries

If we assume translational invariance in OPS, then Eq. (1) gives

$$\dot{x} + \dot{\varepsilon} = F_\lambda(x + \varepsilon) \tag{7}$$

for ε infinitesimal. Taylor-expanding the r.h.s. about a particular point $x = \xi$ and using (1), we get

$$\dot{\varepsilon} = F'_\lambda(\xi).\varepsilon \tag{8}$$

where F'_λ is the Jacobian matrix. Assuming a normal mode solution,

$$\varepsilon = \varepsilon_0 \exp(\omega_t t) \tag{9}$$

Eq. (8) gives

$$\omega_t \varepsilon_0 = F'_\lambda(\xi).\varepsilon_0 \ . \tag{10}$$

The translational normal mode exponents ω_t are the roots of the characteristic equation:

$$|F'_\lambda(\xi) - 1\ \omega_t(\xi,\lambda)| = 0 \ . \tag{11}$$

When $\xi = x_c$, (11) coincides with the criterion employed in customary linear stability analysis. In the present context, it is the consequence of a postulated translation symmetry in OPS. The modes $\omega_t(\xi,\lambda)$ in general correspond to transients. In the light of the criteria spelt out earlier, if there is at least one $\omega_t(\xi,\lambda)$ which is ≥ 0, we say that translation invariance is spontaneously broken at $x = \xi$.

A rotation in an N-dimensional OPS is defined by

$$x \to x' : \ x' = Rx \ , \tag{12}$$

where R belongs to O(N); for an infinitesimal rotation

$$R = 1 + \delta \tag{13}$$

where δ is an antisymmetric matrix of infinitesimals $(\delta_1, \delta_2, \ldots, \delta_p)$ with $p = N(N-1)/2$.

Imposing rotational invariance for the system in OPS, we have

$$\dot{x}' = F_\lambda(x') \ . \tag{14}$$

For a reference point $x = \xi$, an infinitesimal rotation gives

$$\dot{\xi} + \dot{\delta}\xi + \delta\dot{\xi} = F_\lambda(\xi + \delta\xi) \ . \tag{15}$$

Using the rotational normal mode ansatz,

$$\delta = \delta_0 \exp(\omega_r t) ,$$

where δ_0 is an antisymmetric constant matrix, we find

$$\omega_r \delta_0 \xi + \delta_0 F_\lambda(\xi) = F'_\lambda(\xi) \delta_0 \xi . \tag{16}$$

At a critical point x_c this equation reduces to

$$\omega_r \delta_0 x_c = F'_\lambda(x_c) \delta_0 x_c . \tag{17}$$

Comparing this with (10) for $\xi = x_c$, it follows that

$$\omega_r(x_c) = \omega_t(x_c) \text{ and } \delta_0(x_c) = \varepsilon_0 \tag{18}$$

modulo some normalisation constant. It is clear that at a critical point translational and rotational symmetries coexist provided the corresponding common ω values are all negative. However, at any other point the two symmetries need not coexist.

As (16) constitutes a system of linear homogeneous equations in the δ_0 unknowns, a consistency condition can be written for the existence of nontrivial solutions. This turns out to be an algebraic equation for ω_r. Contrary to expectations, when $N \geq 2$ the degree of this equation is only N-1. This is apparently a consequence of the antisymmetry of the δ_0 matrix, although we are unable to offer a general proof at the moment. Just as in the case of translational invariance we say that at any point $x = \xi$ in OPS rotational invariance is spontaneously broken if at least one of the ω_r values is ≥ 0.

3. Scale Symmetry

For a dissipative system with an N dimensional OPS we define a static (i.e., time independent) scale transformation by

$$x_i \to \Sigma x_i$$

$$F_i \to \Sigma^{\Delta_i} F_i(\Sigma x) \qquad (i = 1,2,\ldots,N) \tag{19}$$

where Σ is a constant scalar scale parameter and the dependence of F_i on the control parameter λ is suppressed; Δ_i is scaling index for the function F_i. In general $\Delta_i = \Delta_i(x)$. In passing we remark that (19) is tailored along the lines of (2) rather than (4).

An infinitesimal static scale transformation is taken in the form $\Sigma = (1 + \sigma)$ where σ is infinitesimal. Then we have

$$\dot{x}_i = (1 + \sigma)^{\Delta_i - 1} F_i[(1 + \sigma)x]$$

or

$$\dot{x}_i = [1 + (\Delta_i - 1)\sigma]F_i(x + \sigma x) \ . \tag{20}$$

For a given point $x = \xi$, we obtain to $O(\sigma)$,

$$\Delta_i(\xi) = 1 - \sum_j F_{ij}(\xi)\xi_j/F_i \tag{21}$$

where $F_{ij}(\xi) = \dfrac{\partial F_i(\xi)}{\partial \xi_j}$.

It may be noted that we have not assumed any homogeneity properties for the F_i.

In the same spirit as above we define a time dependent or dynamic scale transformation as follows:

$$x \longrightarrow \Sigma(t)x$$

$$F(x) \to \Sigma^{\Delta}(t)F(\Sigma(t)x) \ . \tag{22}$$

Here $\Sigma(t)$ is a time dependent matrix of order N and Δ is a (scalar) scaling index which is independent of time. By assumption Δ is introduced for the whole system of N differential equations. In our infinitesimal case let us assume that

$$\Sigma(t) = 1 + \sigma(t) \tag{23}$$

where $\sigma(t)$ is a matrix of infinitesimal components. Then the dynamically scaled equations of motion read

$$\frac{d}{dt}[(1 + \sigma(t))x] = [1 + \sigma(t)]^{\Delta} F_\lambda[(1 + \sigma(t))x] \ . \tag{24}$$

Replacing $[1 + \sigma(t)]^{\Delta}$ by $1 + \Delta\sigma(t)$ and choosing a specific point $x = \xi$

$$\dot{\sigma}\xi + \sigma F_\lambda(\xi) = \Delta\sigma\, F_\lambda(\xi) + F'_\lambda(\xi)\sigma\xi \ . \tag{25}$$

To determine the local stability of a dynamically scaled trajectory passing through $x = \xi$, we use the normal mode ansatz:

$$\sigma(t) = \sigma_0 \exp(\omega_s t) \ , \tag{26}$$

leading to the equation

$$\omega_s\sigma_0\xi + (1 - \Delta)\sigma_0 F_\lambda(\xi) = F'_\lambda(\xi)\sigma_0(\xi) \ . \tag{27}$$

At a critical point this becomes

$$F'_\lambda(x)\sigma_0 x_c = \omega_s \sigma_0 x_c \ . \tag{28}$$

This equation defines the same eigenvalue problem as considered in the preceding section. Hence (18) can be extended to include dynamic scaling also:

$$\varepsilon_0 = \delta_0 x_c = \sigma_0 x_c \tag{29}$$

upto normalizations. It follows trivially that σ_0 tends to be an antisymmetric matrix δ_0 (modulo normalization) as $x \to x_c$. Thus we find that all the three symmetries, namely, translation, rotation and dynamic scaling, coexist at critical points, the latter two invariances being almost indistinguishable as a critical point is approached.

The definition of dynamic scaling given in (22) is not a generalization of the static scaling considered on the basis of (19). It is certainly possible to generalize (19) by letting Δ_i as well as σ become time-dependent:

$$\begin{aligned} x_i &\to [1 + \sigma(t)]x_i \\ F_i &\to [1 + \sigma(t)]^{\Delta_i(t)} F_i[(1 + \sigma(t))x] \end{aligned} \tag{30}$$

Using the normal mode assumption $\sigma(t) = \sigma_0 \exp(\omega_s t)$, the equations of motion yield the relation

$$\omega_s \xi_i = (\Delta_i - 1)F_i(\xi) + \sum_j F_{ij}\xi_j \ . \tag{31}$$

If the static approximation for Δ_i is invoked ω_s works out to be just zero, as it should. A non-trivial ω_s would arise, only if $\Delta_i = \Delta_i(t)$. At this point a distinction between the two types of dynamic scaling is in order. Dynamic scaling defined by (22) is called Type I scaling, characterised as it is by $\Delta =$ constant, and that defined by (30) is called Type II scaling which is associated with N time dependent scaling indices $\Delta_i(t)$.

Both Type I and Type II have the common feature that the scaling indices act like additional control parameters. Thus for Type I, $\omega_s = \omega_s(\xi;\lambda,\Delta)$ and for Type II, $\omega_s = \omega_s(\xi;\lambda, \Delta_1,\ldots,\Delta_N)$. As an infinite number of Δ or Δ_i values can be selected at each point, ω_s cannot be uniquely fixed; indeed, an infinite number of ω_s values with arbitrary signs can be introduced at every point. This implies that dynamic scale invariance cannot be broken anywhere in OPS. Even if the set of ω_s values turns out to be positive or zero, an equal number of negative ω_s values may be picked up at the same point, thanks to the arbitrariness of the Δ or Δ_i values.

4. Pattern of Spontaneous Symmetry Breaking in the Lorenz Model

The Lorenz model [5], as briefly referred to in the Introduction, originally arose as a three mode truncation of the Navier-Stokes equation for fluid convection. It has since been found to describe varied physical situations such as irregular spiking in lasers [6] and the dynamics of weakly unstable finite amplitude baroclinic waves [7]. We have chosen this model to test the concepts herein introduced mainly because it is the most well studied system in the literature and as such, there exists a wealth of numerical data to guide us. It is also typical of different kinds of chaotic behaviour such as preturbulence, intermittency, period doubling strange attractor,etc.

The Lorenz equations are

$$\begin{aligned} \dot{x}_1 &= -\sigma x_1 + \sigma x_2 \\ \dot{x}_2 &= -x_1 x_3 + \gamma x_1 - x_2 \\ \dot{x}_3 &= x_1 x_2 - b x_3 . \end{aligned} \tag{32}$$

The system was studied [5] at $\sigma = 10$, $b = 8/3$ and by varying γ . Here we summarise the behaviour observed for various values of γ. For $0 < \gamma < 1$ the origin is the only critical point which is globally attracting. For $\gamma > 1$ this point becomes unstable and two new critical points

$$x_{c\ 1,2} = \{\pm(b(\gamma - 1))^{1/2}, \pm(b(\gamma - 1))^{1/2}, \gamma - 1\} \tag{33}$$

appear; these are attracting for $1 < \gamma < \gamma_2$ where $\gamma_2 = \sigma(\sigma + b+3)/(\sigma-b-1)$ which is 24.74, for the above choice of σ and b. For $\gamma > \gamma_2$ all the critical points are unstable and no simple attractors exist in the system. At $\gamma = \gamma_2$ a Hopf bifurcation occurs. But before this, around $\gamma = \gamma_1 = 13.926....$, trajectories starting near the origin cross over to the fixed point on the other side. As γ increases beyond γ_1, a strange invariant set is found to exist in the $(\gamma-1)$ plane and this set exhibits divergence of nearby trajectories, a phenomenon known as preturbulence or metastable chaos. As γ is increased further, this set moves in OPS and at $\gamma = \gamma_3 (= 28)$, a strange attractor emerges. The development of the strange attractor reverses itself for $\gamma > 28$, and periodic limit cycles are found to occur for some higher γ values.

We have studied translational symmetry in the OPS of the Lorenz model by computing ω_t values with $b = 8/3$ and $\sigma = 10$ and by varying the parameter γ. It is found that at every point one of the three

ω_t values is always negative and independent of the control parameter γ. This presumably points to the dissipative nature of the system being reflected in a tendency for global contraction of the phase space. The nature of the other two ω_t values determines stability against translational symmetry breaking. Thus the origin is found to be stable for $\gamma < 1$. At $\gamma = 1$ one of the ω_t values vanishes and remains positive beyond $\gamma = 1$, in agreement with the available stability information from the literature. So when the fixed point becomes unstable translational symmetry in OPS is broken.

The stability analysis is carried out not only for these fixed points, x_{c1} and x_{c2} , but for a large number of points in the x_1-x_2 plane for different values of x_3. We give the outcome of one such investigation in Fig. 1. The shaded region corresponds to stable points. It is interesting to note that beyond $\gamma = 1$, the structure of these regions is independent of γ. It is the fixed points which do depend on γ that move from the stable into the unstable region as γ approaches 24.74.

The normal mode exponents for rotational symmetry are calculated for the Lorenz model. Although most of the numerically available results refer to the x_1-x_2 plane with $x_3 = \gamma - 1$, we find there is nothing sacred about this plane especially for high γ-values. Moreover in the present case when $x_3 = \gamma - 1$, equation (16) gets reduced to a linear equation in ω_r and so there is only one relevant ω_r in this plane. To be a little more general, we proceed with our analysis in an arbitrary x_1-x_2 plane. Here again, for high values of γ, say γ=500, where only stable limit cycles exist the pattern in which stable and unstable regions occur, remains the same for any value of x_3. For

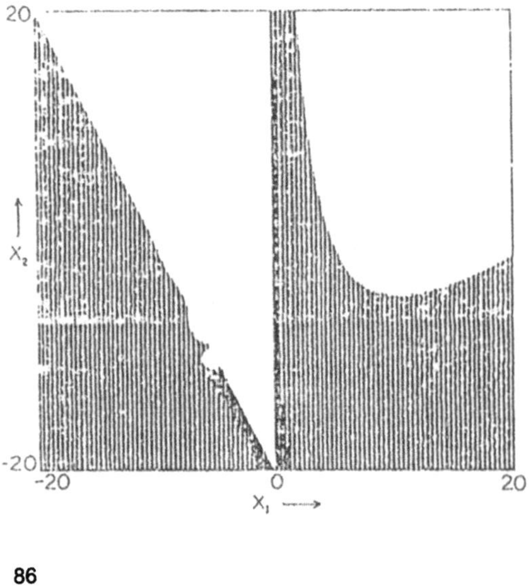

Fig. 1 Stable and unstable regions in the x_1-x_2 plane for translational symmetry. The shaded regions correspond to stable regions.

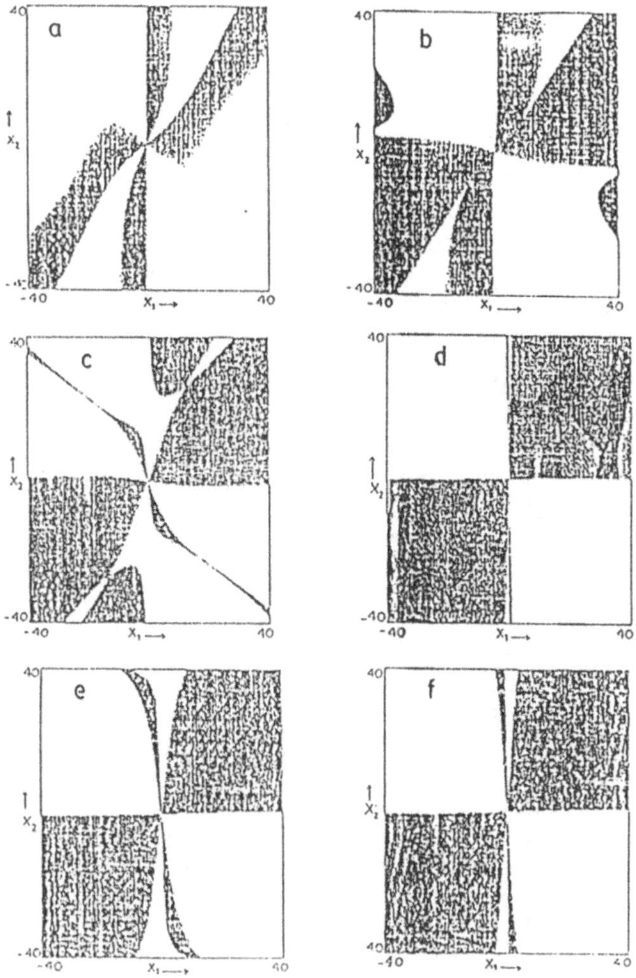

Fig. 2 Patterns showing rotationally stable and unstable regions in the x_1-x_2 plane for different values of γ. Here x_3 = 27.
a) γ = 13 b) γ = 28 c) γ = 50 d) γ = 500
e) γ = 99.529 f) γ = 166.3

smaller values of γ this pattern depends on the choice of the x_1-x_2 plane, i.e., on the value that we fix for x_3. Generally as $x_3 \approx \gamma$ more stable regions are found to develop symmetrically and thereafter, there is a slow evolution of the structure of stable and unstable regions as γ keeps changing. In Fig. 2, we show this change for a few chosen values of γ. However, before and after period doubling or in the region near intermittency, there is no marked or sudden change in the pattern. This is clear from Figs. 2e-f. The boundary separating stable and unstable regions becomes more and more well defined as γ is increased.

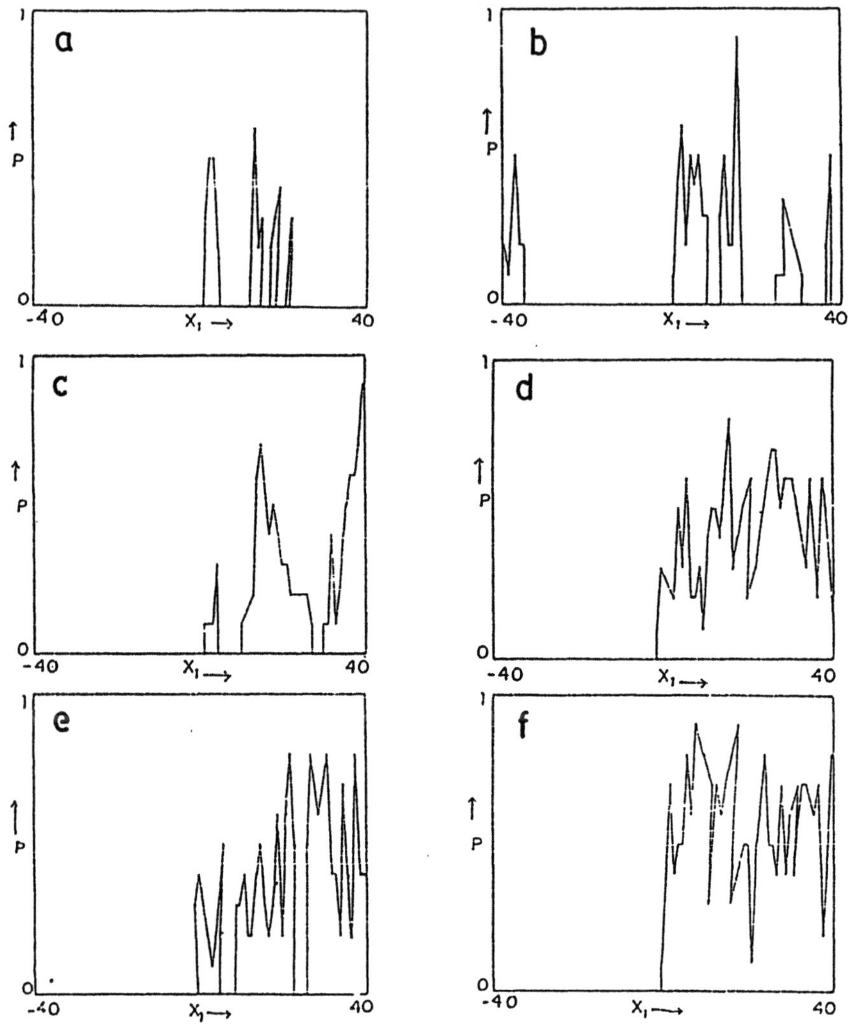

Fig. 3 The variation of P along the x_1 direction with $x_3 = 27$ and $x_2 = 10$ and for chosen values of γ: a) $\gamma = 13$, b) $\gamma = 28$, c) $\gamma = 500$, d) $\gamma = 100.5$, e) $\gamma = 99.98$, f) $\gamma = 166.3$

To express the above evolution in the pattern more quantitatively, we have computed the number N_s of stable points in a unit square that give rise to negative values of ω_r. This number divided by the total number of points N considered within the square, i.e., $P = N_s/N$ represents the probability for the occurrence of stable points. The variation of P with x_1 for $x_2 = 10$ and $x_3 = 27$ is plotted in Fig. 3 for different values. These pictures clearly indicate that there is structure inside the stable regions also. For small values of γ, where preturbulence and chaos exist, the variation of P depends crucially on the selection of the x_1-x_2 plane. However, for γ values where stable limit cycles exist, we find that horizontal regions are present in the curve of P vs x_1 indicating decrease in the fluctuations of P (e.g. γ = 50, 100.5, 500, etc.). A comparison of the curves for γ = 100.5 and γ = 99.5255 should make this clear. At γ = 100.5, stable one cycle is known to exist, while γ = 99.5255 corresponds to the accumulation point of period doubling bifurcations. We also note this in the region near γ = 166.3, where intermittent behaviour is observed. Thus rapid fluctuations in P from point to point in OPS do not favour the formation of stable cycles.

At this stage, it is not obvious whether the rotational symmetry analysis gives much more additional information than what is obtained from translational symmetry analysis. However we feel, the formation of the stable limit cycles as well as their symmetry properties will depend to a great extent on the pattern of stable and unstable regions and the fluctuations in P. Further investigations on other similar model systems are necessary before general conclusions can be drawn.

5. Concluding Remarks

In the present work we have carried out an analysis of continuous symmetries of dissipative systems in OPS. We have demonstrated the possibility of employing a normal mode exponent as the local indicator of stability against the spontaneous breaking of a particular symmetry. Thus if the exponent is zero or positive at a point in OPS, the given point is unstable and the particular symmetry at that point is spontaneously broken. If a finite region of OPS is found to exhibit a large degree of SSB, it may be associated with mild chaos or preturbulence.

As our study of Lorenz model indicates, even in the region where a strange attractor is known to exist the spontaneous breaking of translation and rotation invariances is not complete. Scale symmetry has been shown to behave differently from the other two symmetries in that two types of dynamical scale invariance and one type of static scale invariance have been defined. All the three types are essential for understanding the vagaries of scale invariance at least at a modest level. Even though other definitions of dynamical scale symmetry cannot be completely ruled out, the most interesting aspect of this kind of invariance is that it cannot be totally broken even locally. If carefully pursued, this result might provide a clue to understanding the self-similar structure found in strange attractors which is probably a manifestation of unbroken scale symmetry.

Acknowledgements

We thank Dr V. M. Nandakumaran for several useful discussions related to the subject. G. A. thanks the CSIR, New Delhi for financial assistance through a research associateship and K. M. V. thanks the Cochin University of Science and Technology for the award of a Junior Research Fellowship.

References

[1] H. Benard, Rev. Gen. Sci. Pure Appl. **11** (1900) 1261.
[2] E. L. Koschmieder in **Order and Fluctuations in Equilibrium and Nonequilibrium Statistical Mechanics**, ed. G. Nicolis, G. Dewel and J. W. Turner (Wiley, New York, 1981), p. 159.
[3] J. S. Nicolis, Rep. Prog. Phy. **49** (1986) 1109.
[4] H. Haken, Rep. Prog. Phys. **52** (1989) 515.
[5] C. Sparrow, **The Lorenz Equations: Bifurcations, Chaos and Strange Attractors** (Springer-Verlag, New York, 1982).
[6] H. Haken, Phys. Lett. **A53** (1975) 77.
[7] J. Pedlosky and C. Frenzen, J. Atmos. Sci. **37** (1980) 1177.

Part II

Singularity Structure Aspects

Painlevé Property in Hamiltonian and Non-Hamiltonian Systems

S. Puri

School of Physical Sciences, Jawaharlal Nehru University,
New Delhi 110067, India

The Painlevé property (viz., the absence of movable singularities other than simple poles in the general solution of a differential equation) has been proposed as a means of determining whether or not a dynamical system is completely integrable. This paper reviews some of the extant work in this area.

1. Introduction

Qualitatively speaking, the motion of a given dynamical system can be regular or irregular. There is a deep connection between the qualitative nature of the motion and the complete integrability of the dynamical system. (We will elaborate the notion of complete integrability shortly.) Loosely speaking, a completely integrable system will exhibit regular motion. To date, there has been no method of identifying whether or not a system is completely integrable other than by explicit demonstration (which can be difficult). However, recent work [1-3] demonstrates a connection between the complete integrability of a dynamical system and the Painlevé property, viz., the absence of movable singularities other than simple poles [4] in the general solution (in complex time) of the equations of motion of the dynamical system. In this paper, we attempt to systematize some of the extant results in this field of research, taking special care to delineate the rigorous results (of which there are few) from the conjectures (of which there are many).

This paper is organised as follows. In Section 2, we define complete integrability in the context of finite-dimensional and infinite-dimensional (∞ - d) Hamiltonian systems. (These are well-established concepts [5-8].) Subsequently, we explain the Painlevé property and the proposed "Painlevé test of integrability" [1-3] for Hamiltonian systems. (In this paper, we do not provide a tutorial on how to apply the "Painlevé test" [which involves doing the singularity analysis of the equations of motion of the system] as this has already been done by a number of authors [1-4].) The concept of complete integrability is relevant in the context of non-Hamiltonian systems also but is not well established for such systems. Following Bountis et al. [9], we provide a working definition of complete integrability for non-Hamiltonian systems. We then explore the connection between complete integrability of non-Hamiltonian systems and the Painlevé property. Section 3 is devoted to a brief summary and discussion of this paper.

2. The Concept of Complete Integrability and the Painlevé Property

The concept of complete integrability is relevant to both Hamiltonian and non-Hamiltonian systems. We discuss it first in the context of Hamiltonian systems, where the idea is more firmly established.

2.1 Hamiltonian systems

Finite-dimensional Hamiltonian systems. A finite-dimensional (with dimension 2N) Hamiltonian system is defined as a system labelled by coordinates $\{q_i\}_{i=1,N}$; their conjugate momenta $\{p_i\}_{i=1,N}$; and a Hamiltonian $H(\{p_i\}_{i=1,N}, \{q_i\}_{i=1,N}, t)$, where t is the time. The equations of motion are derivable from the Hamiltonian as

$$\dot{q}_i = \frac{\partial H}{\partial p_i} ,$$

$$\dot{p}_i = - \frac{\partial H}{\partial q_i} , \qquad (2.1)$$

where the dot denotes differentiation with respect to time and i ranges from 1 to N. The Hamiltonian system is said to be completely integrable [7,8] if it has N constants of motion $\{K_i(p,q)\}_{i=1,N}$ (which are globally defined single valued functions of their arguments) such that

$$\dot{K}_i(p,q) = 0 \text{ for all } i \; ; \qquad (2.2)$$

and these constants are in involution, viz., given any K_i, K_j their Poisson bracket is zero:

$$\left\{K_i, K_j\right\} = \sum_{k=1}^{N} \left[\frac{\partial K_i}{\partial q_k} \frac{\partial K_j}{\partial p_k} - \frac{\partial K_i}{\partial p_k} \frac{\partial K_j}{\partial q_k} \right]$$

$$= 0 \text{ for all } i,j. \qquad (2.3)$$

Then, the motion is on an N-dimensional manifold in phase space. The Liouville-Arnol'd theorem [8] states that, for a completely integrable finite-dimensional Hamiltonian system (whose motion is on a compact manifold in phase space), there exists a canonical transformation which takes the system into action-angle variables $\{\omega_i\}_{i=1,N}$ and $\{J_i\}_{i=1,N}$ with equations of motion

$$\dot{J}_i = 0 \text{ or } J_i = a_i ,$$

$$\dot{\omega}_i = b_i \text{ or } \omega_i = b_i t + c_i \text{ for all } i, \tag{2.4}$$

where a_i, b_i and c_i are constants of integration. This corresponds to quasiperiodic motion on an N-torus. By this stage we have explictly solved the equations of motion, so we are justified in using the term "integrable" to describe the system. Examples of finite-dimensional completely integrable Hamiltonian systems are the harmonic oscillator and the two-body central force problem. Notice that a Hamiltonian system can be explicitly solvable (by quadratures) without being completely integrable. The requirement of complete integrability is a stronger requirement than that of solvability.

(∞ - d) Hamiltonian systems. We can generalize some of the above concepts to the case with infinite dimensions [5]. Consider an ∞-d dynamical system, which is labelled by generalized coordinates $q(x,t)$ (where x is a spatial coordinate and t is time) and their conjugate generalized momenta $p(x,t)$. It is assumed that $p(x,t)$ and $q(x,t)$ are analytic functions of x on $|x| <\infty$ (so that all order spatial derivatives are defined). The system is said to be Hamiltonian if the equations of motion are derivable from a Hamiltonian $H[p,q,t]$ (which is a functional of $p(x,t)$ and $q(x,t)$ and their spatial derivatives) as

$$q_t = \frac{\delta H}{\delta p} ,$$

$$p_t = -\frac{\delta H}{\delta q} . \tag{2.5}$$

An example of such a system in 1-space is

$$q_t = iq_{xx} - 2iq^2 p ,$$

$$p_t = -ip_{xx} + 2iqp^2 . \tag{2.6}$$

This is derivable from the Hamiltonian

$$H[p,q,t] = -i \int_{-\infty}^{\infty} [q_x p_x + q^2 p^2] dx , \tag{2.7}$$

where $q(x,t)$, $p(x,t)$, $q_x(x,t)$ and $p_x(x,t)$ are assumed to decay sufficiently rapidly as $|x| \to \infty$. If we consider the special case $p(x,t) =$

$\pm$ $q^*(x,t)$ (complex conjugate of $q(x,t)$), then (2.6) reduces to the Nonlinear Schrödinger equation

$$q_t = iq_{xx} \pm 2i|q|^2 q. \tag{2.8}$$

The corresponding Hamiltonian is

$$H[q^*,q,t] = -i \int_{-\infty}^{\infty} [\pm q_x^2 + |q|^4] dx . \tag{2.9}$$

We can also define the Poisson bracket as

$$\{A,B\} = \int_{-\infty}^{\infty} \left\{ \frac{\delta A}{\delta q}\frac{\delta B}{\delta p} - \frac{\delta A}{\delta p}\frac{\delta B}{\delta q} \right\} dx , \tag{2.10}$$

where A, B are functionals of $p(x,t)$ and $q(x,t)$ which decay sufficiently rapidly as $|x| \to \infty$. Two functionals are said to be in involution if their Poisson bracket is zero.

A transformation from (p,q) to (P,Q) is said to be a canonical transformation if

$$\{Q(x,t),\ Q(y,t)\} = 0 ,$$

$$\{P(x,t),\ P(y,t)\} = 0 ,$$

$$\{Q(x,t),\ P(y,t)\} = \delta(x - y) . \tag{2.11}$$

There is no ∞-d analog to the Liouville-Arnol'd theorem. Thus, it is not useful to define complete integrability in terms of constants of motion. Instead, Ablowitz and Segur [5] directly define (in analogy with the finite-dimensional case) an ∞-d Hamiltonian system to be completely integrable if there is a canonical transformation which reduces the system to a system of action-angle variables, viz., we can find a canonical transformation $(p,q) \to (J,\omega)$ such that

$$J_t(x,t) = 0 \quad \text{or} \quad J(x,t) = A(x) ,$$

$$\omega_t(x,t) = B(x) \quad \text{or} \quad \omega(x,t) = B(x)t + C(x) , \tag{2.12}$$

where $A(x)$, $B(x)$ and $C(x)$ are continuous functions of x.

This completes our definition of complete integrability in finite dimensional and ∞-d Hamiltonian systems.

2.2 *"Painlevé Test" for Hamiltonian Systems*

Kowalevskaya [10] was the first to recognize a connection between the complete integrability of a finite-dimensional Hamiltonian system and the analytic structure of its equations of motion in the complex plane. The problem she considered was that of the motion of a rigid body about a fixed point. She performed the apparently unphysical calculation of finding the parameter values for which the general solution of the equations of motion admitted no movable (initial condition dependent) branch-points or essential singularities in the complex plane. For all such values she was able to solve the equations of motion explicitly. For other parameter values the solutions are as yet unknown [11]. Subsequently, Painlevé and Gambier [12] catalogued all first and second order differential equations (not necessarily derivable from a Hamiltonian) which had the property that their only movable singularities in the complex plane were simple poles. This later came to be known as the Painlevé property [4]. All but six of these equations had already known convergent series solutions. The solutions of these six equations were called the Painlevé transcendents.

"Painlevé Test" of Ablowitz et al. [1]. Interest in this work was revived when Ablowitz et al. [1] discovered that the Painlevé transcendents arose as solutions to ordinary differential equations obtained by considering reductions of partial differential equations (**derivable from a Hamiltonian**)[1] which were solvable by the Inverse Scattering Transform (a specialized technique for solving completely integrable ∞-d Hamiltonian systems [5,6]). They proposed the so-called "Painlevé test" for partial differential equations (**derivable from a Hamiltonian**): If all reductions (perhaps after a change of variables) of a partial differential equation (**derivable from a Hamiltonian**) to an ordinary differential equation have the Painlevé property then the original partial differential equation is solvable by the Inverse Scattering Transform.[2]

Some examples of "Painlevé test" of Ablowitz et al. [1]. Insofar as it is possible to consider all possible reductions of a partial differential equation to an ordinary differential equation, all partial differential equations which pass the "Painlevé test" of Ablowitz et

---

[1]Though they do not state this explicitly, Ablowitz et al. [1] consider only Hamiltonian systems and the test was presumably proposed only in the context of Hamiltonian systems. Subsequently, the test came to have meaning in the context of non-Hamiltonian systems also.

[2]Given the historical background of Kowalevskaya's work, one may argue that it would be more logical to propose that passing the "Painlevé test" implies complete integrability. However, solvability of an ∞-d Hamiltonian system by the Inverse Scattering Transform is equivalent to complete integrability [5,6].

al. [1] are solvable by the Inverse Scattering Transform. A number of examples have been provided by Ablowtiz et al. [1].

The converse, however, is somewhat trickier. There are examples of partial differential equations that are solvable by the Inverse Scattering Transform but do not appear to pass the "Painlevé test" of Ablowitz et al. [1], viz., the Benjamin-Ono equation [13], the intermediate long-wave equation [14]. These are integrodifferential equations and, because of the nonlocality of the integral kernel, a local singularity analysis cannot be performed. Grammaticos et al. [15] have rewritten these equations in a form in which they are seen to pass the "Painlevé test" of Ablowitz et al. [1]. We are not aware of any other examples of equations solvable by the Inverse Scattering Transform which do not pass the "Painlevé test" of Ablowitz et al. [1].

"Painlevé test" of Weiss et al. [2]. Clearly, it is impossible to consider all possible reductions of a partial differential equation, because these are infinite in number. Thus, Weiss et al. [2] have proposed a generalized version of this test, which is directly applicable to partial differential equations.

Some examples of "Painlevé test" of Weiss et al. [2]. As far as we know, all partial differential equations which pass the "Painlevé test" of Weiss et al. [2] are solvable by the Inverse Scattering Transform. Weiss et al. [2] consider some well-known equations which pass the "Painlevé test" proposed by them and are solvable by the Inverse Scattering Transform. Apart from these examples, there are a number of other field equations which pass the "Painlevé test" of Weiss et al. [2] and are solvable by the Inverse Scattering Transform, e.g., Chudnovsky et al. [16] study multicomponent isospectral deformation equations, Ovel and Steeb [17a] consider a time-dependent Kadomtsev-Petviashvili equation [17b], Ramani et al. [18a] consider the Hirota-Satsuma equations [18b] (which are expected to be solvable by the Inverse Scattering Transform, though this has yet to be shown), and Grauel [19a] considers the Maxwell-Bloch equations [19b].

There is, however, an example of an equation which can be solved by the Inverse Scattering Transform but does not pass the "Painlevé test" of Weiss et al. [2], viz., the Harry Dym equation $u_t = u^3 u_{xxx}$ [2b,20]. No generalization has yet been proposed which can include this equation in the purview of the "Painlevé test". Thus, empirical evidence suggests that the passing of the "Painlevé test" of Weiss et al. [2] is sufficient but no necessary condition for a partial differential equation to be solvable by the Inverse Scattering Transform.

Some examples of "Painlevé test" of Kowalevskaya [10]. The "Painlevé test" in the form originally used by Kowalevskaya has also been applied directly to finite-dimensional Hamiltonian systems. Again, there do not seem to be any examples of systems which pass the "Painlevé test" of Kowalevskaya [10] but are not completely integrable. Using it, Bountis et al. [21] were able to identify the completely integrable cases for the Hénon-Heiles system [22] and for a system of coupled anharmonic oscillators. They provide explict solutions in these cases. The test has also been used by Bountis et al. [21] and Adler and van Moerbeke [23] to identify completely integrable cases for the Toda lattice [24]. Lakshmanan and Sahadevan [25] have identified some of the completely integrable cases for systems of N coupled quartic anharmonic oscillators using a weaker form of the "Painlevé test" proposed by Ramani et al. [26a] (which we motivate shortly). However, their analysis was incomplete and has been completed by Ramani et al. [26b; see also 26c].

There are a number of examples [26a] of completely integrable systems which do not pass the "Painlevé test" of Kowalevskaya. To account for these, Ramani et al. [26a] have proposed a "weak Painlevé test" which accounts for all the known counterexamples.

Rigorous results for Hamiltonian systems. There is no analytic proof of a general connection between the Painlevé property and complete integrability of a Hamiltonian system. Kozlov [27a] and Ziglin [27b] have independently considered certain simple Hamiltonian systems (e.g., a heavy rigid body moving about a fixed point, the Hénon-Heiles system) and shown that there is no meromorphic first integral of the motion **except** in the cases where these systems possess the Painlevé property. Also, for a class of completely integrable Hamiltonians (called algebraically completely integrable Hamiltonians) Adler and van Moerbeke [23] have shown that all solutions must necessarily possess the Painlevé property. The sufficiency condition is not yet proven, though it is expected to hold [9]. The precise connection between algebraic complete integrability and the Painlevé property has been explained by Yoshida [27c]. These seem to be the only rigorous studies available in the case of Hamiltonian systems.

2.3 Non-Hamiltonian Systems

These systems are not derivable from a Hamiltonian, unlike in the previous case. Thus the notion of complete integrability is not clear. The methods of singularity analysis can still be applied, though, and can be used to give a working definition of complete integrability [9].

Finite-dimensional non-Hamiltonian systems. Various studies of singularities in finite-dimensional non-Hamiltonian systems suggest the following working definition [9]: A non-Hamiltonian system (of N dimensions) may be said to be completely integrable, if

(i) it can be reduced to a final quadrature by the existence of (N-1) integrals of motion; or

(ii) it can be transformed (by a change of variables) into a set of linear ordinary differential equations (with variable coefficients); or

(iii) it can be reduced to an equation with the Painlevé property.

In the finite-dimensional case, one may expect that completely integrable systems (in the above sense) show regular motion [9]. This is analogous to the case of completely integrable finite-dimensional Hamiltonian systems.

(∞-d) Non-Hamiltonian systems. For ∞-d non-Hamiltonian systems, the definition is even trickier. Perhaps the "safest" definition of complete integrability in ∞-d non-Hamiltonian systems is in analogy with the finite-dimensional case, suggesting the following working definition: An ∞-d non-Hamiltonian system may be said to be completely integrable if

(i) it can be transformed (by a change of variables) into a linear partial differential equation; or

(ii) it can be reduced to an equation which passes the "Painlevé test" for partial differential equations [2].

We should emphasize that there is no clear connection between complete integrability of ∞-d non-Hamiltonian systems (defined in this fashion) and regular motion of the system. Thus, Burgers' equation [28] passes the "Painlevé test" for partial differential equations [2] and is (by the second clause of our working definition) completely integrable. However, it exhibits highly complicated trajectories. On the other hand, the d.c. driven, damped sine-Gordon equation [29] does not pass the "Painlevé test" for partial differential equations [30] and is not linearizable. Thus, it is not completely integrable by our working definition. Nevertheless, it exhibits highly regular motion [31].

2.4 *"Painlevé Test" for Non-Hamiltonian Systems*

Some examples of "Painlevé test" of Kowalevskaya [10]. By our working definition, finite-dimensional systems which pass the "Painlevé test" of Kowalevskaya [10] are completely integrable. By itself, this would not be particularly useful. However, various studies (quoted below) show that if a system passes the "Painlevé test" [10], it can be solved (in principle). Segur [32] has performed the singularity analysis for the case of the Lorenz equations (which are non-Hamiltonian) [33]. Segur was able to identify four completely integrable cases where he could (in principle) solve the equations of motion. Bountis et al. [9,34a] and Steeb et al. [34b] have considered the singularity structure of the Lotka-Volterra equations [34c]. These authors have identified completely integrable cases for these equations and have proceeded to solve them for these cases. Finally, we mention three more studies of singularities in finite-dimensional non-Hamiltonian systems. Bountis et al. [9] have used the "Painlevé test" of Kowalevskaya [10] to identify (and integrate) completely integrable cases for the three-wave interaction problem [35a] and the Rikitake dynamo model [35b]. Menyuk et al. [36] have applied singularity analysis to identify completely integrable cases for the general multiple-wave interaction problem.

We are not aware of any examples in which completely integrable finite-dimensional systems (by either of the first two clauses of our working definition) do not possess the Painlevé property.

Some examples of "Painlevé test" of Weiss et al. [2]. Again, by our working definition, an ∞-d system that passes the "Painlevé test" of Weiss et al. [2] is completely integrable. Weiss et al. [2] have shown that Burgers' equation (our second example) passes the "Painlevé test" they propose for partial differential equations. It is well known that Burgers' equation can be linearized into the diffusion equation by means of the Cole-Hopf transform [37] and, therefore, is completely integrable by the first clause of our working definition also. (However, it is not clear what this says about the behaviour of solutions of Burgers' equation as these are known to exhibit complicated trajectories rather than any particularly regular motion.)

There is an example of an equation (viz., the nonlinear diffusion equation $u_t = (u^{-2}u_x)_x$) which can be linearized into the diffusion equation (and is, therefore, completely integrable by the first clause of our working definition) but does not pass the "Painlevé test" of Weiss et al. [2]. As yet, there has not been a generalization of the "Painlevé test" [2] which can account for this counterexample.

Rigorous results for non-Hamiltonian systems. Again, there is a paucity of rigorous work. Anderson et al. [38] have considered systems of ordinary differential equations which have nonlinear superposition principles; well-defined Lie-algebraic structures; and are linearizable. These are shown to have the Painlevé property.

3. Summary and Discussion

In this paper, we discussed the connection between the complete integrability of (Hamiltonian and non-Hamiltonian) dynamical systems and the Painlevé property. The connection appears to be a profound one but has not yet been established analytically. Most conclusions in this field are based on the strength of empirical evidence.

In Section 2, we defined the concepts of complete integrability and the Painlevé property. The concept of complete integrability is well established for finite-dimensional and ∞-d Hamiltonian systems. For non-Hamiltonian systems, the definition of complete integrability is not yet well established. Following Bountis et al. [9], we have presented a working definition of complete integrability for finite-dimensional and ∞-d non-Hamiltonian systems. This definition is suggested by the singularity analysis of non-Hamiltonian systems. We discuss the Painlevé property and the proposed "Painlevé test" of Kowalevskaya [10], Ablowitz et al. [1] and Weiss et al. [2]. Empirical evidence suggests that passing the "Painlevé test" is sufficient but not necessary condition for a dynamical system to be completely integrable. From the examples we have provided for (finite-dimensional and ∞-d) Hamiltonian systems and (finite-dimensional) non-Hamiltonian systems, it is clear that the "Painlevé test" [1,2,10] is a useful means of identifying completely integrable systems, which (in turn) are solvable in principle and exhibit regular motion. For ∞-d non-Hamiltonian systems, considerably more examples need to be considered before we can evaluate the utility of the "Painlevé test" [2]. However, the examples of non-Hamiltonian partial differential equations considered by us make it clear that there is no connection between our definition of complete integrability and regular motion in these systems. Perhpas the major obstacle to further progress is the scarcity of analytical results. We have mentioned some extant analytical results but these are restricted to special classes of systems.

REFERENCES

[1] M. J. Ablowitz, A. Ramani and H. Segur, Lett. Nuovo Cim. **23** (1978) 333; M. J. Ablowitz, A. Ramani and H. Segur, J. Math. Phys. **21** (1980) 715, 1006.

[2a] J. Weiss, M. Tabor and G. Carnevale, J. Math. Phys. **24** (1983) 522.

[2b] J. Weiss, J. Math. Phys. **24** (1983) 1405; **25** (1984) 13.

[3a] A. Ramani, B. Grammaticos and T. Bountis, Phys. Rep. **180** (1989) 159; and references therein.
[3b] S. Puri, to appear in Intl. J. Mod. Phys. B. (1990).
[4] E. L. Ince, **Ordinary Differential Equations** (Dover, New York, 1956); and references therein.
[5] M. J. Ablowitz and H. Segur, **Solitons and the Inverse Scattering Transform,** SIAM Studies in Applied Mathematics 4 (SIAM Press, Philadelphia, 1981); and references therein.
[6] A. C. Newell, **Solitons in Mathematics and Physics**, CBMS-NSF Regional Conference Series in Applied Mathematics 48 (SIAM Press, Philadelphia, 1985); and references therein.
[7] J. Moser, **Stable and Random Motions in Dynamical Systems**, Ann. Math. Studies 77 (Princeton University Press, Princeton, 1973).
[8] V. I. Arnol'd, **Mathematical Methods of Classical Mechanics,** Graduate Texts in Mathematics 60 (Springer-Verlag, New York, 1978).
[9] T. C. Bountis, in **Singularities and Dynamical Systems**, ed. S. N. Pnevmatikos (North-Holland, Amsterdam, 1984); T. C. Bountis, A. Ramani, B. Grammaticos and B. Dorizzi, Physica **A** **128** (1984) 268.
[10] S. Kowalevskaya, Acta Math. **12** (1889)177;**14**(1889) 81.
[11] V. V. Golubev, **Lectures on Integrals of Motion** (State Publishing House, Moscow, 1953).
[12a] P. Painlevé, Bull. Soc. Math. France **28** (1900) 227; C. R. Acad. Sc. Paris **135** (1902) 411,641,757,1020.
[12b] B. Gambier, C. R. Acad. Sc. Paris **142** (1906) 1403,1497.
[13] T. B. Benjamin, J. Fluid Mech. **29** (1967) 559; H. Ono, J. Phys. Soc. Jpn. **39** (1975) 1082.
[14] G. B. Whitham, Proc. Roy. Soc. London **A299** (1967) 6; R. I. Joseph, J. Phys. **A10** (1977) L225.
[15] B. Grammaticos, B. Dorizzi and A. Ramani, Phys. Rev. Lett. **53** (1984) 1.
[16] D. V. Chudnovsky, G. V. Chudnovsky and M. Tabor, Phys. Lett. **A97** (1983) 268.
[17a] W. Oevel and W.-H. Steeb, Phys. Lett. **A103** (1984) 239.
[17b] B. B. Kadomstev and V. I. Petviashvili, Dokl. Akad. Nauk SSSR **192** (1970) 753.
[18a] A. Ramani, B. Dorizzi and B. Grammaticos, Phys. Lett. **A99** (1983) 411.
[18b] R. Hirota and J. Satsuma, Phys. Lett. **A85** (1981) 407.
[19a] A. Grauel, J. Phys. **A19** (1986) 479.
[19b] J. C. Eilbeck, J. D. Gibbon, P. J. Caudrey and R. K. Bullough, J. Phys. **A6** (1973) 1337.
[20] K. M. Case, in A. I. P. Conference Proceedings, **Mathematical Methods in Hydrodynamics and Related Dynamical Systems,** December 1981, La Jolla, ed. M. Tabor and Y. M. Treve (American Institute of Physics, New York, 1982).
[21] T. C. Bountis, H. Segur and F. Vivaldi, Phys. Rev. **A25** (1982) 1257.
[22] M. Hénon and C. Heiles, Astron. J. **69** (1964) 73.
[23] M. Adler and P. van Moerbeke, Commun. Math. Phys. **83** (1982) 83.
[24] M. Toda, **Theory of Nonlinear Lattices,** Springer-Verlag Series in Solid State Sciences 20 (Springer-Verlag, New York, 1981).
[25] M. Lakshmanan and R. Sahadevan, Phys. Lett. **A103** (1984) 189; Phys. Rev. **A31** (1985) 861.
[26a] A. Ramani, B. Dorizzi and B. Grammaticos, Phys. Rev. Lett. **49** (1982) 1539; B. Dorizzi, B. Grammaticos and A. Ramani, J. Math. Phys. **24** (1983) 2282; B. Grammaticos, B. Dorizzi and A. Ramani, J. Math. Phys. **24** (1983) 2289.
[26b] A. Ramani, J. Hietarinta, B. Dorizzi and B. Grammaticos, Phys. Lett. **A108** (1985) 55.

[26c] R. Sahadevan and M. Lakshmanan, Phys. Rev. **A33** (1986) 3563.
[27a] V. V. Kozlov, Prikt. Mat. Mekh. **42** (1978) 400.
[27b] S. L. Ziglin, Funkts. Anal. Prilozhen **16** (1982) 30; **17** (1983) 8.
[27c] H. Yoshida, Cel. Mech. **31** (1983) 363, 381.
[28] J. M. Burgers, Trans. Roy. Neth. Acad. Sci. Amsterdam **17** (1939) 1; Adv. Appl. Mech. **1** (1948) 171.
[29] D. W. Mclaughlin and A. C. Scott, Phys. Rev. **A18** (1978) 1652.
[30] S. Puri, Phys. Lett. **A107** (1985) 359.
[31] S. Puri and Y. Oono, unpublished.
[32] H. Segur, Lectures at International School of Physics "Enrico Fermi", Varenna, Italy (1980).
[33] E. N. Lorenz, J. Atmos. Sci. **20** (1963) 130.
[34a] T. C. Bountis, M. Bier and J. Hijmans, Phys. Lett. **A97** (1983) 11.
[34b] W.-H. Steeb, A. Kunick and W. Strampp, J. Phys. Soc. Japan **52** (1983) 2649.
[34c] N. S. Goel, S. C. Maitra and E. W. Montroll, Rev. Mod. Phys. **43** (1971) 231.
[35a] A. S. Pikovskii and M. I. Ravinovich, Math. Phys. Rev. **2**, Sov. Sci. Rev. C, Harwood (1981).
[35b] T. Rikitake, Proc. Cambridge Phys. Soc. **54** (1957) 89.
[36] C. R. Menyuk, H. H. Chen and Y. C. Lee, Phys. Rev. **A27** (1983) 1597; J. Math. Phys. **24** (1983) 1073.
[37] E. Hopf, Comm. Pure Appl. Math. **3** (1950) 201; J. D. Cole, Quart. Appl. Math. **9** (1951) 225.
[38] R. Anderson, J. Harnad and P. Winternitz, Physica **D4** (1982) 164.

Singularity Structure and Chaotic Dynamics of the Parametrically Driven Duffing Oscillator

S. Parthasarathy

Department of Physics, Bharathidasan University,
Tiruchirapalli 620024, India

The nonintegrability aspects of the generalised parametrically driven Duffing oscillator is reviewed by investigating analytically the singularity structure exhibited by the solution of the system in the complex time domain. The simultaneous contraction and rotation of this singularity pattern in the $z = t^4$ lnt plane, as the control parameter is varied, results in complicated clustering of singularities in the t-plane is pointed out. The corresponding chaotic dynamics of the system is studied numerically.

1. Introduction

Singularity structure aspects of the solutions in the complex time plane of nonintegrable nonlinear dynamical systems have received considerable attention in recent times [1-7]. The standard singularity structure or Painlevé (P) analysis [8,9] is concerned with the investigation of the conditions under which the solution of a given differential equation is meromorphic, exhibiting movable singularities of pole type (and their generalizations). Such investigations are useful in identifying the integrable cases of nonlinear dynamical systems [10,11]. The recent investigations [1-7,11] show that P-analysis can also be profitably used as a powerful tool to study the nonintegrability aspects, especially through the analytic structure studies of the solutions. Some of the nonintegrable dynamical systems which have been studied in this way are the Lorenz equation [1,6], Henon-Heiles system [2,3,6], Duffing oscillator [4-6] and so on. It has been proved that in all these systems that for selected parameter ranges the solution exhibits very complicated analytic structure. In certain nonintegrable dynamical systems admitting irrational or complex powers in the local expansions, it was observed [3,5-7] that the singularities lie on spirals about the movable singular point and accumulate and thereby leading to the formation of self-similar natural boundaries in the complex t-plane. Other recent studies [4] reveal that transverse intersections of separatrices necessarily imply the existence of solutions which

Research Reports in Physics
Editors: M. Lakshmanan · M. Daniel

Symmetries and Singularity Structures

are multivalued with infinitely many branches in the complex t-plane and thus indicate the connection between the Melnikov criterion [12] for the onset of chaos and analytic structure of nonintegrable Hamiltonian and non-Hamiltonian systems.

In this article we give a preliminary report on the study of the analytic structure of the generalised parametrically driven Duffing oscillator (DO) based on the work of Levine and Tabor [6] for the standard DO and show numerically its corresponding chaotic dynamics. The Duffing oscillator [13] which models a variety of physical processes such as buckled beam oscillations [14,15], plasma oscillations [16], nonlinear circuits [17] and so on, admits a strange attractor and exhibits [14,15] chaotic behaviour through period-doubling bifurcations. Recently the analytic structure of the DO has been systematically investigated by several authors [4-7,11] and a clustering of singularities in the complex t-plane was observed. We use this methodology [5] to study here the analytic structure of the generalized parametrically driven Duffing oscillator.

2. The Parametrically Driven Duffing Oscillator

We consider the generalised parametrically driven Duffing oscillator of the form

$$\ddot{x} + p\dot{x} + (\omega_0^2 + f_2 \cos\omega_2 t)x + \beta x^3 = f_1 \cos\omega_1 t \ , \tag{1}$$

where dot refers to differentiation with respect to time and the parameters have their usual meaning. To begin with we represent the solution to (1) locally as a Laurent series [3,5] of the form

$$x = \sum_{j=0}^{\infty} a_j \tau^{j-1} \ , \qquad \tau = (t - t_0) \to 0 \tag{2}$$

about a movable singularity t_0, in the complex time (t-) plane. For (2) to be a local representation of the general solution of (1), one of the coefficients a_j in (2) should be arbitrary, in addition to t_0. The recursion relation for a_j's [3,5] can be given in the form

$$a_j(j-1)(j-2) + pa_{j-1}(j-2) + \omega_0^2 a_{j-2} + f_2 \sum_{r=0}^{j} G_{j-r-2}\, a_r + \beta \sum_{r,p} a_{j-r} a_{r-p} a_p = f_1 F_{j-3} \ , \quad 0 \le p \le r \le j \ , \tag{3}$$

where $F(t) = \cos\omega_1 t$, $F_j = \frac{1}{j!} \frac{\partial^j F}{\partial t^j}\Big|_{t=t_0}$ and $G(t) = \cos\omega_2 t$, $G_j = \frac{1}{j!} \frac{\partial^j G}{\partial t^j}\Big|_{t=t_0}$.

The values of first few coefficients in (3) are easily noted to be

$$a_0^2 = -2/\beta \ , \quad a_1 = -\tfrac{1}{6}pa_0 \ , \quad a_2 = \tfrac{1}{36}(6\omega_0^2 - p^2 + 6f_2 \cos\omega_2 t_0)a_0,$$

$$a_3 = -\tfrac{1}{108}pa_0(2p^2 - 9\omega_0^2) - \tfrac{1}{4}f_1 \cos\omega_1 t_0 + \tfrac{1}{12} f_2 a_0(p \cos\omega_2 t_0 - 3\omega_2 \sin\omega_2 t_0) \tag{4}$$

and for a_4 we get

$$(0)a_4 - \tfrac{1}{27}p^2 a_0(2p^2 - 9\omega_0^2) - f_1(p \cos\omega_1 t_0 - \omega_1 \sin\omega_1 t_0)$$
$$+ \tfrac{1}{6} f_2 a_0[(2p^2 - 3\omega_2^2)\cos\omega_2 t_0 - 5p\omega_2 \sin \omega_2 t_0] = 0 \ . \tag{5}$$

From (5) it is evident that a_4 is the needed arbitrary coefficient. But evidently this cannot be true for arbitrary values of the parameters and t_0, which necessiates the introduction of logarithmic terms in Laurent series thereby making the solution be of non P-type. However, for the special choice of the parameters $f_1 = 0$ and $f_2 = 0$, two possibilities arise from (5) for a_4 to be arbitrary, viz. (i) $p = 0$, which implies that the equation of motion of the free undamped oscillator is of meromorphic type which in fact corresponds to the Jacobian elliptic function solutions [18,19], and (ii) $p = \pm \frac{3}{\sqrt{2}} \omega_0$, which implies the equation of motion of the free damped oscillator possesses P-property for these two specific choices of p. We have recently presented [20] their corresponding exact solutions.

3. The Logarithmic psi series

For the general case we introduce the psi series [5,6],

$$x = \sum_{j=0}^{\infty} \sum_{k=0}^{\infty} a_{jk}\tau^{j-1}(\tau^4 \ln\tau)^k \ . \tag{6}$$

Then the recursion relations for the a_{jk} for (1) become

$$a_{jk}[(j+4k-1)(j+4k-2)] + a_{j-4,k+1}(k+1)(2j+8k-3)$$
$$+ a_{j-8,k+2}(k+1)(k+2) + pa_{j-1,k}(j+4k-2) + pa_{j-5,k+1}(k+1)$$
$$+ \omega_0^2 a_{j-1,k} + f_2 \sum_{r=0}^{j} G_{j-r-2}a_{r,k} \ \delta_{0k}$$
$$+\beta \sum_{\substack{p,q\\ r,s}} a_{j-r,k-s}\, a_{r-p,s-q}\, a_{pq} = f_1 F_{j-3} \ \delta_{0k} \ ,$$

$$0 \leq p \leq r \leq j, \quad 0 \leq q \leq s \leq k \tag{7}$$

and the values of the coefficients a_{00}, a_{10}, a_{20} and a_{30} take on the same values as given by (4) for a_0, a_1, a_2 and a_3 respectively. For a_{40} to be arbitrary we now have

$$(0)a_{40} + 5a_{01} - \frac{1}{27} p^2 a_{00}(2p^2 - 9\omega_0^2) - f_1(p \cos\omega_1 t_0 - \omega_1 \sin\omega_1 t_0)$$

$$+ \frac{1}{6} f_2 a_{00}[(2p^2 - 3\omega_2^2)\cos\omega_2 t_0 - 5p\omega_2 \sin\omega_2 t_0] = 0 , \tag{8}$$

which means that if

$$a_{01} = \frac{1}{135} p^2 a_{00}(2p^2 - 9\omega_0^2) + \frac{1}{5} f_1(p \cos\omega_1 t_0 - \omega_1 \sin\omega_1 t_0)$$

$$- \frac{1}{30} f_2 a_{00}[(2p^2 - 3\omega_2^2)\cos\omega_2 t_0 - 5p\omega_2 \sin\omega_2 t_0] \tag{9}$$

then a_{40} is arbitrary. Thus t_0 and a_{40} are the two pieces of arbitrary data entering into (6). Evidently (6) signals the presence of a movable logarithmic branch point and (1) is, in general, nonintegrable except when (i) $f_1 = 0$, $f_2 = 0$; $p = 0$ and (ii) $f_1 = 0$, $f_2 = 0$; $p = \pm \frac{3}{\sqrt{2}} \omega_0$.

In order to study the analytic structure of the solution of (1), we now look for a closed set of recursion relations amongst a_{jk}'s. These turn out to be the set a_{0k}, $k = 0,1,2,\ldots$ which satisfy

$$16k(k - 1)a_{0k} + 4ka_{0k} + 2a_{0k} + \beta \sum_{s,q} a_{0,k-s}\, a_{0,s-q}\, a_{0q} = 0 . \tag{10}$$

4. The Analytic Structure

Introducing now the generating function

$$\theta(z) = \sum_{k=0}^{\infty} a_{0k} z^k , \tag{11}$$

where z is some (as yet unspecified) independent variable, the following differential equation for $\theta(z)$ is obtained:

$$16z^2\theta'' + 4z\theta' + 2\theta + \beta\theta^3 = 0 , \tag{12}$$

where prime denotes differentiation with respect to z. Since in the limit $\tau \to 0$, the most dominant terms in the psi series (6) involve powers of $\tau^4 \ln \tau$ only, we can obtain (12) in a more direct method by substituting

$$x = \frac{1}{\tau} \Theta(z) , \tag{13}$$

where

$$z = \tau^4 \ln \tau \tag{14}$$

into (1). Thus (12) can be regarded as the original equation of motion (1) **rescaled** in the neighbourhood of a given singularity.

We can also see that (12) can be integrated exactly in terms of elliptic functions by making the substitution

$$\Theta(z) = y f(y) , \quad y = z^{\frac{1}{4}} \tag{15}$$

in (12), gives rise to

$$f''(y) + \beta f^3(y) = 0 , \tag{16}$$

where prime refers to differentiation with respect to y. Integration of (16) once gives

$$(f')^2 + \frac{\beta}{2} f^4 = I_1 , \tag{17}$$

where the value of the integration constant can be determined as $I_1 = -i\ 10\sqrt{2}(a_{01}/\sqrt{\beta})$. By a simple scaling transformation $f(y) = (2I_1/\beta)^{\frac{1}{4}}\ g[(\beta I_1/2)^{\frac{1}{4}} y]$, (17) is reduced to the standard form

$$g'^2(\tilde{y}) + g^4(\tilde{y}) = 1 , \quad \tilde{y} = (\beta I_1/2)^{\frac{1}{4}} y . \tag{18}$$

Equation (18) can be readily integrated and its solution is given by

$$g(\tilde{y}) = \pm i\sqrt{2}\ \mathrm{ds}[\sqrt{2}(\tilde{y} - \tilde{y}_0)] , \tag{19}$$

where $\tilde{y}_0$ is some initial phase and the modulus of the elliptic function $k = \frac{1}{2}$. Choosing $\tilde{y}_0 = 0$, we can write

$$f(y) = i \frac{2}{3^{3/4}\sqrt{\beta}} \xi^{\frac{1}{4}}\ \mathrm{ds}\left(\frac{\sqrt{2}}{3^{3/4}} \xi^{\frac{1}{4}}\ y\right) , \tag{20}$$

where

$$\xi = 2p^2(2p^2 - 9\omega_0^2) - 9f_2[(2p^2 - 3\omega_2^2)\cos\omega_2 t_0 - 5p\omega_2 \sin\omega_2 t_0]$$
$$- i\ 27\sqrt{2}\ f_1\sqrt{\beta}(p \cos\omega_1 t_0 - \omega_1 \sin\omega_1 t_0) .$$

The corresponding pole positions labelled by the lattice site integers

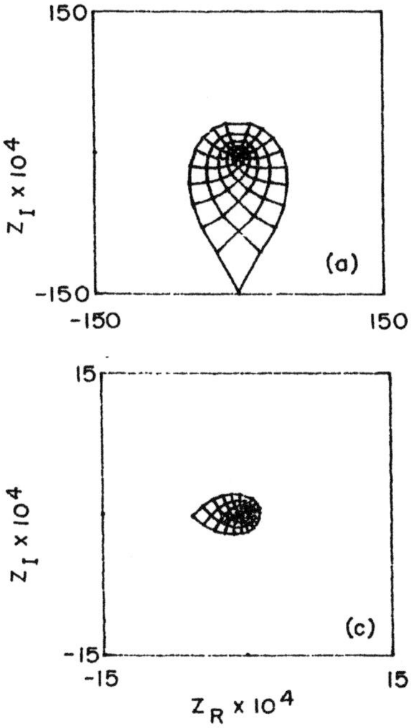

Fig. 1 Singularity structure of Eq. (1) in the complex z-plane ($z = t^4 \ln t$) for $\omega_0^2 = 0.01$, $p = 3\omega_0/\sqrt{2}$, $\beta = 10.0$, $f_1 = 1.5$, $\omega_1 = 1.0$, $\omega_2 = 1.5$, $\ell, m = -10, -9, \ldots, -1, 0, 1, \ldots, 9, 10$ and (a) $f_2 = 0.0$, (b) $f_2 = 3.5$, (c) $f_2 = 15.0$

ℓ, m are given by [18,19]

$$y_{\ell m} = (108)^{\frac{1}{4}} K(\tfrac{1}{2}) \xi^{-\frac{1}{4}} (\ell + im) , \quad \ell, m \in Z . \tag{21}$$

The singularity positions in the z-plane can be obtained from the pole positions $y_{\ell m}$ as

$$z_{\ell m} = 108[K(\tfrac{1}{2})]^4 \frac{(AC - BD) + i(AD + BC)}{A^2 + B^2} , \tag{22}$$

where

$$A = 2p^2(2p^2 - 9\omega_0^2) - 9f_2[(2p^2 - 3\omega_2^2)\cos\omega_2 t_0 - 5p\omega_2 \sin\omega_2 t_0] ,$$
$$B = 27\sqrt{2}\, f_1 \sqrt{\beta}(p \cos\omega_1 t_0 - \omega_1 \sin\omega_1 t_0) ,$$
$$C = (\ell^2 - m^2)^2 - 4\ell^2 m^2 ,$$
$$D = 4\ell m(\ell^2 - m^2).$$

From (22) we can study the singularity structure in the complex z-plane by plotting z_I versus z_R for a chosen set of parametric values. We have fixed the parameter values as $\omega_0^2 = 0.01$, $p = 3\omega_0/\sqrt{2}$, $\beta = 10.0$, $f_1 = 1.5$, $\omega_1 = 1.0$, $\omega_2 = 1.5$ and obtained the singularity structure in the complex z-plane about the singularity located at the origin (since $t_0 = 0$) for various f_2 values. When $f_2 = 0.0$ the singularity structure is as shown in Fig. 1a, where the singularities lie at

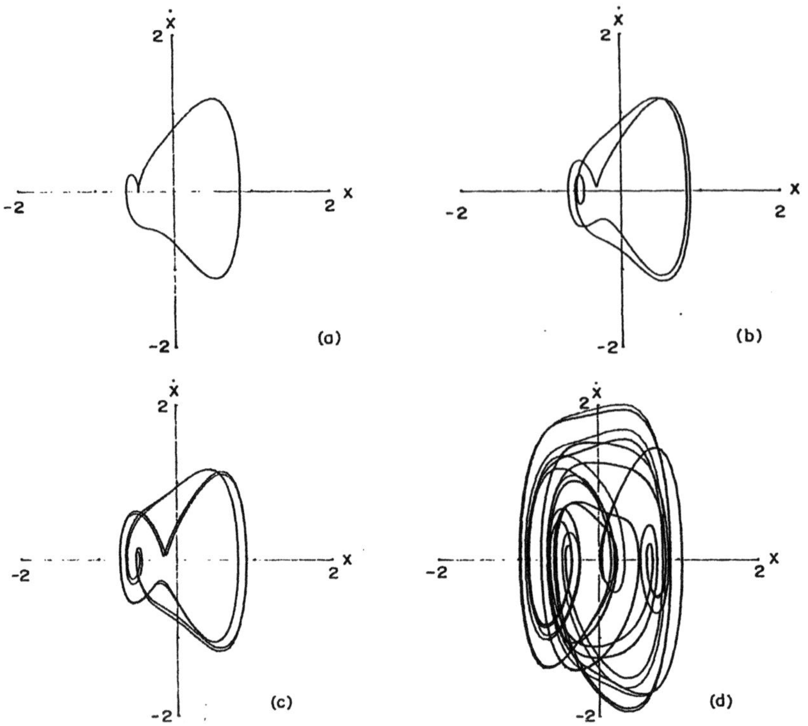

Fig. 2 Phase portraits (x-$\dot{x}$ plots) showing the period-doubling phenomenon of the system (1) with $\omega_0^2 = 0.01$, $p = 3\ \omega_0/\sqrt{2}$, $\beta = 10.0$, $f_1 = 1.5$, $\omega_1 = 1.0$, $\omega_2 = 1.5$ and (a) $f_2 = 0.0$, period T limit cycle, (b) $f_2 = 0.5$, period 2T limit cycle, (c) $f_2 = 1.25$, period 4T limit cycle, (d) $f_2 = 3.5$ chaotic trajectories.

those points at which the two lines cross each other. For $f_2 = 3.5$ the singularity structure undergoes a contraction and rotation which is depicted in Fig. 1b. Further increase in the value of f_2, say 15.0 makes the singularity structure undergo further contraction and rotation as shown in Fig. 1c and thus leads to a very complicated structure. It is to be noted that for each of the singularity in the complex z-plane we can map [5] numerically the corresponding singularity and its substructure in the complex t-plane by using polar coordinates in both the z and t-planes. Thus an immensely complicated multisheeted structure in the complex t-plane can be envisaged.

In order to understand the dynamics in the real time domain corresponding to the above structure, we have numerically integrated the equation of motion (1) by keeping the above mentioned parameter values fixed and varying f_2 in steps. It is obvious that for $f_2 = 0.0$ one gets a limit cycle of period T in the phase plane ($\dot{x}$ vs x) as shown in Fig. 2a. As f_2 is increased further the system exhibits period doubling bifurcation route to chaos. For f_2 values 0.5 and 1.25 we get period 2T and 4T limit cycles (cf. Fig. 2b and 2c) respectively. For $f_2 = 3.5$ we get the phase portrait showing chaotic oscillations (Fig. 2d).

5. Conclusion

To conclude, the study of analytic structure of singularities of solutions of nonintegrable nonlinear dynamical systems provides valuable information about the dynamics. In the above study we have made a preliminary analysis of the singularity structure exhibited by the solution of the parametrically driven Duffing oscillator and presented its corresponding chaotic dynamics. A more detailed report will be published elsewhere.

Acknowledgment

I wish to thank Professor M. Lakshmanan for asking me to investigate this problem and offering his guidance.

References

[1] M. Tabor and J. Weiss, Phys. Rev. **A24** (1981) 2157.
[2] T. Bountis, H. Segur and M. Vivaldi, Phys. Rev. **A25** (1982) 1257.
[3] Y. F. Chang, J. M. Greene, M. Tabor and J. Weiss, Physica **8D** (1983) 183.
[4] T. Bountis, V. Papageorgiou and M. Bier, Physica **24D** (1987) 292.
[5] J. D. Fournier, G. Levine and M. Tabor, J. Phys. **A21** (1988) 33.
[6] G. Levine and M. Tabor, Physica **33D** (1988) 189.
[7] M. Tabor, Pramana J. Phys. **33** (1989) 315.
[8] M. J. Ablowitz, A. Ramani and H. Segur, J. Math. Phys. **21** (1980) 715.
[9] M. Lakshmanan and R. Sahadevan, Phys. Rev. **A31** (1985) 861.
[10] M. Lakshmanan and R. Sahadevan, "Painlevé Analysis, Lie Symmetries and Integrability of Coupled Nonlinear Oscillators of Polynomial Type," Phys. Rep., to appear (1990).
[11] A. Ramani, B. Grammaticos and T. Bountis, Phys. Rep. **180** (1989) 160.
[12] V. K. Melnikov, Trans. Moscow Math. Soc. **12** (1963) 1.
[13] G. Duffing, **Erzwnungene Schwingungen bei veränderlicher Eigenfrequenz und ihre technische Bedeutung** (Vieweg, Braundschweig, 1918).
[14] J. Guckenheimer and P. Holmes, **Nonlinear Oscillations, Dynamical Systems, and Bifurcations of Vector Fields** (Springer, Berlin, 1983).
[15] F. C. Moon, **Chaotic Vibrations** (John Wiley, New York, 1987).
[16] R. A. Mahaffey, Phys. Fluids **19** (1976) 1387.
[17] Y. H. Kao, J. C. Huang and Y. S. Gou, Phys. Rev. **A35** (1987)5228.
[18] H. T. Davis, **Introduction to Nonlinear Differential and Integral Equations** (Dover, New York, 1962).
[19] M. Abramowitz and I. A. Stegun, **Handbook of Mathematical Functions** (Dover, New York, 1965).
[20] S. Parthasarathy and M. Lakshmanan, J. Sound & Vibration **137** (1990) 523.

A Singularity Analysis Approach to the Solutions of Duffing's Equation

T. Bountis [1], *M. Bier* [2;*], *and V. Papageorgiou* [2]

[1]Department of Mathematics, University of Patras, 261 10 Patras, Greece
[2]Department of Mathematics and Computer Science, Clarkson University, Potsdam, NY 13676, USA
* Present address: Department of Mathematics, Colby College, Waterville, ME 04901, US

The singularity structure of Duffing's equation in the complex t-plane is investigated analytically and numerically. A series expansion for the general solution around each singularity $t_* = t_R + it_I$ *is given, and is subsequently used to approximate the locations of singularity "spirals"* $t_*^{(n)}$, $n = 1,2,\ldots$, *around every* t_*. *The main "chimney" patterns--on which singularities are observed to accumulate--are explained by deriving a simple expression for the distances between singularities on the "walls" of these chimneys* $|t_*-t_*^{(1)}| \sim Q^{-1/4}\exp(-t_I/2)$, *Q being the amplitude of the (periodic) driving force. Thus, singularity patterns are seen to further "condense", as Q increases and the motion becomes* **globally** *more chaotic in real t. These results suggest that series expansions near singularities in the complex t-plane can provide useful representations of the general solution of Duffing's equation.*

In recent years, there has been considerable progress in the understanding of the global behaviour of the solutions of nonlinear dynamical systems, from the viewpoint of their singularity analysis in the complex time plane [1-4]. Most of the activity so far has focused on completely integrable systems [5-7], many new examples of which have been discovered by the requirement that their solutions possess the Painlevé property, i.e., that their only movable singularities in the complex t-plane are poles.

More recently, however, with the help of some important theorems due to Ziglin [8,9], it has been realized that the singularity analysis of **non-integrable** systems can be put on a more rigorous footing, as well. In particular, the work of Yoshida [10] and others [11,12] has shown that it is possible to prove **non-integrability** of 2-degree of freedom Hamiltonian systems by examining their singularity structure. On the other hand, the splitting of separatrices has been rigorously connected with multi-valued, infinitely sheeted solutions, even in the case of non-Hamiltonian, near-integrable systems [13]. For a more recent review of the results of singularity analysis to date, see [14].

Research Reports in Physics Symmetries and Singularity Structures
Editors: M. Lakshmanan · M. Daniel

In this paper, we develop the singularity analysis of non-integrable systems further, by demonstrating the usefulness of series expansions of the solutions near their singularities in the complex t-plane. As an example, we have chosen to study here the perennial paradigm of non-integrable dynamical systems, Duffing's equation,

$$\ddot{x} + \delta\dot{x} + Ax + x^3 = Q \cos 2t \tag{1}$$

whose solutions, near a singularity t_*, are of the form [13]

$$x(t) = \frac{1}{\tau} \sum_{r=0}^{\infty} \sum_{j=0}^{\infty} \alpha_{rj} \tau^r (\tau^4 \ln\tau)^j, \quad \tau \equiv t - t_*, \quad \tau \to 0. \tag{2}$$

Here, we will set $\delta = 0$ in (1) but entirely similar results hold in the dissipative case ($\delta > 0$) as well [15].

It has already been demonstrated numerically that the movable singularities t_* of the solutions of (1) lie on chimney patterns, which have smaller subchimneys of singularities extending outward from the "walls" of the primary chimneys [14,15]. Using asymptotic reductions of the solution (2), and concentrating mainly on its leading r= 0 part, we shall demonstrate here analytically that the distance ρ_1, between two neighbouring singularities t_* and t_*', on the walls of the main chimneys, behaves (far enough from the Ret axis) as

$$\rho_1 \sim Q^{-1/4} \exp(-|\mathrm{Im}t_*|/2) .$$

This explains both the exponential accumulation of t_* on these chimneys as well as their increased "condensation", observed numerically as Q increases (see Fig. 1). Moreover, our results also explain

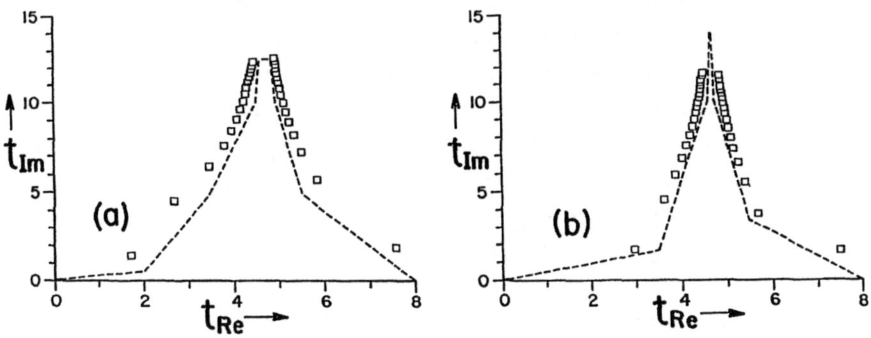

Fig. 1 Chimney patterns of singularities of Duffing's equation (1) for A = -1, δ = 0.25 and (a) Q = 0.2, (b) Q = 0.4 (taken from ref. 13). Note the relative "condensation" of the pattern from (a) to (b). The picture is very similar for δ = 0.

the presence of subchimneys, by demonstrating that around each t_* there is a **spiral** of singularities at distances $\rho_1 > \rho_2 > \rho_3 > \dots$ lying on different Riemann sheets.

Consider, for simplicity, the conservative Duffing's equation

$$\ddot{x} + Ax + x^3 + Q\cos 2t = 0 . \tag{3}$$

When Q = 0, of course, its solutions are well known: They are Jacobi elliptic functions [16], whose singularity patterns consist of a doubly periodic lattice of poles near which the solutions can be expressed as Laurent series

$$x(t) = \frac{1}{\tau} \sum_{\tau=0}^{\infty} \alpha_r \tau^r , \quad \tau \equiv t - t_* , \tag{4}$$

t_* being the (arbitrary) location of these poles determined by the initial conditions for each orbit.

Now, while orbits of (3), for Q = 0, are periodic, and the motion in the x, $\dot{x}$ plane is regular and well understood, the situation for $Q \neq 0$ is completely different: The orbits are free to explore a 3-dimensional phase space, the Hamiltonian of the system is no longer conserved, and no general solution of the equation of motion is analytically available. The x, $\dot{x}$ plane has become a Poincaré surface of section [17], and new, **chaotic** orbits exist, which are seen to intersect the $x,\dot{x}$ surface of section in an apparently irregular and unpredictable manner.

The question we ask here is: Is this interesting behaviour of the solutions reflected to some extent in their singularity patterns in the complex t-plane? And if so, what can we learn from these patterns that will improve our understanding of the motion in **real** t?

To answer this question, note first that these singularities are no longer poles, as in the Q = 0 case (cf. (4)). In fact, near them, the asymptotic expansions of the solutions are known to contain logarithmic terms, as shown in (2), and thus be multivalued functions of t, with generally infinitely many Riemann sheets [9]. All α_{rj} coefficients in (2), are solved, as usual, from recursion relations, obtained by substituting (2) in (3), except for α_{40}, which is the **second** free constant of the problem specified by the initial conditions (the first one is t_*) [13].

To analyze the structure of these chimney patterns, we shall apply the method of **asymptotic reductions**, which has been used successfully in the singularity analysis of other non-integrable dynamical systems [2,3,15]. In the case of Duffing's equation one starts by writing (2) in the form

$$x(t) = \frac{1}{\tau} \sum_{r=0}^{\infty} \tau^r f_r(u), \qquad u \equiv \tau^4 \ln\tau \tag{5}$$

where the functions $f_r(u)$ are assumed to be analytic near $u = 0$. They are, of course, expected to contain--as r increases--terms of smaller and smaller magnitude near $u = 0$. By this reasoning, we can consider the first term in (5), i.e., $x(t) \simeq f_0/\tau$, as the **lowest order** approximation to the solution (2), containing the most divergent terms as $\tau \to 0$. Substituting (5) in (3), and equating terms of the same order in τ, yields a nonlinear ODE for $f_0(u)$,

$$16u^2 f_0''(u) + 4uf_0'(u) + 2f_0(u) + f_0^3(u) = 0 , \tag{6}$$

and a hierarchy of **linear** ODEs for the higher order approximants $f_r(u)$:

$$16u^2 f_r'' + 4(2r + 1)uf_r' + (r^2 - 3r + 2 + 3f_0^2)f_r = F_r(f_0,\dots,f_{r-1}), \tag{7}$$

where

$$F_1 = 0, \quad F_2 = -Af_0, \quad F_3 = -Q\cos 2t_* , \text{ etc.}$$

Now, (6) turns out to be analytically solvable! Making the transformation

$$f_0 = \nu\Theta_0(\nu), \qquad \nu \equiv u^{1/4} \tag{8}$$

one obtains from (6):

$$\Theta_0''(\nu) + \Theta_0^3(\nu) = 0$$

which is easily solved in terms of Jacobi elliptic functions to furnish

$$f_0(u) = -\lambda u^{1/4}\, \mathrm{cn}(\lambda u^{\frac{1}{4}} + iK, 1/\sqrt{2}\,) \tag{9}$$

where $K = K(1/\sqrt{2})$ is the elliptic integral of the first kind. In fact, the more general transformation

$$f_r(u) = \nu^{1-r}\Theta_r(\nu), \qquad u \equiv u^{1/4}, \quad r \geq 0 ,$$

simplifies (7) to an (inhomogeneous) equation of the Lamé-type [18]:

$$\Theta_r''(\nu) + 3\Theta_0^2(\nu)\Theta_r(\nu) = \nu^{3-r}F_r , \tag{10}$$

whose homogeneous part can be exactly solved in terms of standard hypergeometric functions in the variable $w = cn^4(\lambda\nu + iK, 1/\sqrt{2})$. Thus, to the extent that the solutions of these linear equations (10) are tractable, one can argue that (2) represents the **general solution** of Duffing's equation near its singularities in the complex t-plane.

As a first attempt to make a quantitative use of these results, let us substitute (2), in (3) and equate like powers of τ, to find

$$\alpha_{00} = f_0(0) = i\sqrt{2}, \quad \alpha_{01} = f_0'(0) = (2Q/5) \sin 2t_*. \tag{11}$$

Combining (11) with (9) gives an equation for λ

$$\lambda^4 = - 4\sqrt{2}Q \exp(-2it_R + 2t_I) , \tag{12}$$

where we have set $t_* = t_R + it_I$ and taken t_I (> 0) large enough, so that the $\exp(-2t_I)$ term in $\sin 2t_*$ can be neglected.

Since $u = 0$, about which $f_0(u)$ is **analytic**, corresponds to $t = t_*$, this means that the singularity u_* of $f_0(u)$ closest to $u = 0$, will correspond to t_*', i.e., the closest singularity to t_* of $x(t)$ in the complex t-plane. And since f_0 is known, from the theory of elliptic functions, the location of u_* is also known:

$$u_* \equiv 16K^4/\lambda^4 = (2\sqrt{2}K^4/Q) \exp(2it_R - 2t_I) \text{ ,(cf. (12))}. \tag{13}$$

Defining then in (13)

$$u_* \equiv \tau_*^4 \ln \tau_* , \quad \tau_* \equiv t_*' - t_* = \rho e^{i\varphi} , \tag{14}$$

we finally arrive at explicit expressions for ρ and φ, for the approximate location of t_*' closest to t_*:

$$\rho^4(\ln^2\rho + \varphi^2)^{1/2} = 2(\sqrt{2}\, K^4/Q) \exp(-2t_I) , \tag{14a}$$

$$4\varphi + \tan^{-1}(\varphi/\ln\rho) = 2t_R . \tag{14b}$$

These results can be used to study the chimney structure of the singularities of Duffing's equation in more detail. However, we can already deduce some useful information from the above analysis: If we denote by ρ_1, φ_1 the solution of (14) with the smallest φ_1 angle, it is not difficult to see that

$$\rho_1 \sim Q^{-1/4} \exp(-t_I/2) , \tag{15}$$

which implies that ρ_1 decreases exponentially in the $\mathrm{Im}\, t > 0$ direction, as the numerically observed chimney patterns of Fig. 1 clearly show.

Perhaps a more important observation about (15) is the dependence of ρ_1 on Q, the amplitude of the driving force. It is well known that, as Q increases, the "size" of the chaotic regions on the surface of section of (3) also increases [15]. What (15) is telling us is that, as one forces the system harder, and the motion becomes **globally** more chaotic, ρ_1 should decrease approximately as $Q^{-1/4}$, and the chimney patterns become more "condensed" as we have indeed observed in earlier publications (see Fig. 1).[1]

In deriving higher order terms in the expansion (5) we have obtained $f_r(u)$, $r = 1,2,3$, in closed form, as particular solutions $g_r(u)$ of (7) in terms of Jacobi elliptic functions [15]. Moreover, we have found $f_4(u) = c_1 f_0'(u) + g_4(u)$, where c_1 is the second free constant of the solution.

Computing, however, from all of the above, higher order corrections to the locations of the nearest singularities t_*', we found that they were **orders of magnitude** smaller than what one gets from (14). In particular, at $r = 4$, τ_* is corrected by an additive term $\sim c_1/(4\ell n\tau_*+1)$ [15]. Since $|\ell n\tau_*| \to \infty$, as $\tau_* \to 0$, this term also turns out to be too small to significantly affect the ρ, φ values obtained from (14), thus demonstrating again the insensitivity of the "upper" part of the chimneys to changes in the initial conditions [13].

We may now use Eq. (14) to determine the location of singularities higher up on the chimneys of the solutions of (3). To this end, we start by assuming that we know the location of one of them, $t_* = t_R + it_I$, say, and solve (14) for ρ, φ, by a simple Newton algorithm, solving first (14a) for φ in terms of ρ and substituting in (14b).

We thus obtain an infinite number of distinct roots $\rho_1 > \rho_2 > \dots$ to which there corresponds a monotonically increasing set of angles $\varphi_1 < \varphi_2 < \dots$. Denoting now by ρ_{th} and φ_{th} the first solution pair of (14), ρ_1 and φ_1, corresponding to the t_*' just above t_* on the main chimney "wall"--we list them in Tables 1 and 2 for several t_*, against their experimentally computed values, ρ_{exp} and φ_{exp}.

Note first that, for both sets of parameter values and initial conditions in Tables 1,2, the theoretical results clearly show, qualitatively, the presence of chimney patterns in the complex t-plane. Quantitatively also, there is satisfactory agreement between theory and experiment, since ρ_{th} and φ_{th} are generally within 10-20% of their experimental values.

---

[1] Note that in [13] this dependence on Q was erroneously printed as Q^{-1}, instead of the correct $Q^{-1/4}$.

Table 1 Q = 0.1; A = 0.2089 : A periodic orbit of (5) of period π

Sing # n above Ret	$\rho_{th}^{(n)}$	$\varphi_{th}^{(n)}$	$\rho_{exp}^{(n)}$	$\varphi_{exp}^{(n)}$
10	0.1490	1.2411	0.141	1.489
11	0.1380	1.2425	0.128	1.493
12	0.1300	1.2437	0.119	1.498
13	0.1207	1.2446	0.110	1.502
14	0.1136	1.2454	0.102	1.505

Table 2 Q = 2.8; A = 0.3 : A chaotic orbit of (5)

Sing. # n above Ret	$\rho_{th}^{(n)}$	$\varphi_{th}^{(n)}$	$\rho_{exp}^{(n)}$	$\varphi_{exp}^{(n)}$
10	0.1480	1.2378	0.140	1.4847
11	0.1370	1.2390	0.127	1.4899
12	0.1279	1.2410	0.118	1.4955
13	0.1198	1.2420	0.109	1.5000
14	0.1129	1.2430	0.101	1.5026

Finally, it is interesting to note that, similar to what has been found for other nonintegrable systems, the singularity patterns of Duffing's equation also have a lot more structure, than what is shown in Fig. 1. Integrating (3) along a path leading outward from one of the main chimenys, secondary chimneys are discovered, which appear to have yet smaller chimneys on them, etc., as seen e.g. in Fig. 2. For

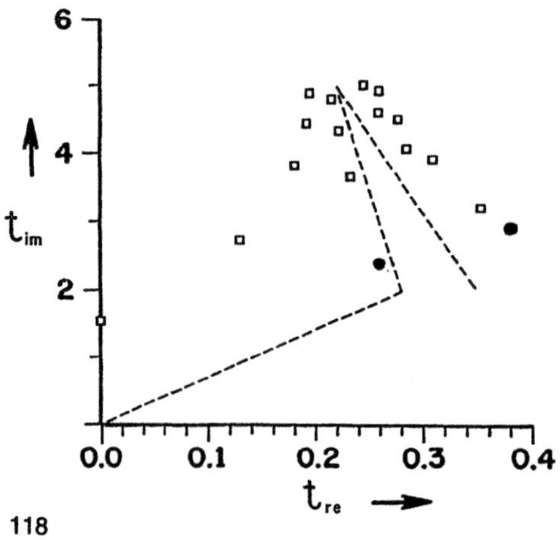

Fig. 2 Secondary chimney singularities marked by □. The dashed line indicates the integration path passing between the second and third singularity (from the left) of Fig. 2(a), marked here by solid dots (●) .

a dtailed analysis of how this "subchimney" structure is described by the roots of (14), ρ_n, φ_n, $n \geq 2$ see ref. [19].

We have presented in this paper, analytical and numerical results on the singularity structure of the solutions of Duffing's equation in the complex t-plane. Our main goal was to use singularity analysis to obtain some useful information about the general solutions of this non-integrable system.

We have shown that the singularities of the solutions of equation (3) lie on "chimney" patterns, on which singularities accumulate densely on the same Riemann sheet. We have studied these patterns by asymptotic reductions of the solutions near their singularities t_* and obtained simple equations for the approximate location of singularities t_*' closest to t_*.

Judging from the accuracy of the above results, it would be reasonable to expect that the series (2) converges over a large region around each t_*, perhaps all the way to the nearest singularities. Thus, one could "patch up" the complex plane with such regions and propose this series as a useful representation of the solution of Duffing's equation, valid especially in places where Taylor series become inaccurate.

Finally, from a more "practical" point of view, it would be very interesting, if relationships such as (15), connecting singularity patterns to global chaotic properties, were to be found in higher dimensional systems as well. One could then look for evidence of global chaos in the accumulation properties of singularity patterns near $t = 0$, rather than integrating many orbits for long times in the Ret direction.

ACKNOWLEDGEMENTS AND A NOTE

Many useful discussions are acknowledged here with C. Cosgrove and M. Tabor. In fact, this paper is a brief review of results obtained already in [15] and reported in [14]. Results very similar to our own on Duffing's equation have also been obtained independently by Fournier, Levine and Tabor (see ref. [19]).

REFERENCES

[1] U. Frisch and R. Morf, Phys. Rev. **A23**(5) (1981) 2673.
[2] M. Tabor and J. Weiss, Phys. Rev. **A24** (1981) 2157.
[3] Y. F. Chang, M. Tabor and J. Weiss, J. Math. Phys. **23**(4) (1982) 531; see also Y. F. Chang et al., Physica **8D** (1983) 183.
[4] T. Bountis in **Singularities and Dynamical Systems**, ed. S. Pnevmatikos (North Holland, Amsterdam, 1985).

[5] A. Ramani, B. Grammaticos and B. Dorizzi, J. Math. Phys. **24** (1983) 2282; see also J. Math. Phys. **25** (1984) 481.
[6] T. Bountis, H. Segur and F. Vivaldi, Phys. Rev. **A25** (1982) 1257.
[7] T. Bountis, A. Ramani, B. Grammaticos and B. Dorizzi, Physica **128A** (1984) 268.
[8] S. L. Ziglin, Funct. Anal. Appl. **16** (1983) 181; also Funct. Anal. Appl. **17** (1983) 6.
[9] S. L. Ziglin, Trans. Moscow Math. Soc. **1** (1982) 283.
[10] H. Yoshida, Celestial Mech. **31** (1983) 363-379, 381-399; see also Physica **21D** (1986) 18.
[11] H. Ito, Kodai Math. J. **8** (1985) 120.
[12] D. Rod, "On an Example of Ziglin in Hamiltonian Dynamics," Conf. Proc. of Canad. Math. Soc. Vol. 8, (AMS, Providence, R.I., 1987).
[13] T. Bountis, V. Papageorgiou and M. Bier, Physica **24D** (1987) 292.
[14] A. Ramani, B. Grammaticos and T. Bountis, "The Painlevé Property and Singularity Analysis of Integrable and Non-Integrable Systems," Phys. Rep. **180**(3) (1989) 160.
[15] M. Bier, Ph.D. Thesis, Department of Mathematics, Clarkson University, Potsdam, NY (1987).
[16] H. T. Davis, **Introduction to Nonlinear Differential and Integral Equations** (Dover, London, 1962).
[17] M. Lieberman and A. Lichtenberg, **Regular and Stochastic Motion** (Springer, Berlin, 1983).
[18] E. T. Copson, **An Introduction to the Theory of Functions of a Complex Variable** (Oxford Press, 1935).
[19] J. D. Fournier, G. Levine and M. Tabor, J. Phys. A: Math. Gen. **21** (1988) 33.

Part III

Integrability and Chaos: Quantum and Classical

Avoided Level Crossing, Solitons and Random Matrix Theory

K. Nakamura

Department of Physics, Fukuoka Institute of Technology,
Higashi-ku, Fukuoka 811-02, Japan

We give a novel insight into mobile avoided crossings observed in the energy spectra of quantum chaos. Exact soliton solutions of generalized Calogero-Moser (gCM) equations are presented. These solutions nicely describe the successive avoided crossings found in the energy-level structure when a nonintegrability parameter is varied. We proceed to study the statistical mechanics of gCM system on the basis of the grand-canonical ensemble. We obtain the probability densities for both the level curvature and level spacing, thereby suggesting an important way to go beyond the traditional framework of the random matrix theory.

1. Miraculous Bridging Between Quantum Chaos and Nonlinear Dynamics

The multiple avoided crossings or repulsions of energy levels are an essential mechanism for complicated spectra with a definite symmetry in quantum bound systems which are classically nonintegrable and chaotic. To understand a statistical aspect of these spectra, considerable attempts have been made to apply the outcome of the random-matrix theory. In particular, much effort has been concentrated upon numerical studies of fitting accumulating data to the level-fluctuation properties predicted by Gaussian orthogonal or unitary ensembles. There is, however, no persuasive and admissible reason why the random-matrix theory, which was exploited by nature in interpreting excitation spectra of heavy nuclei, should be applicable to the analysis of quantum chaos. The most fundamental problem will be to derive complicated level arrangements and wavefunction patterns by investigation of the original quantum Hamiltonian, without having recourse to any ambiguous procedure whose validity is not well proved.

Let us regard τ as a nonintegrability parameter (e.g., strength of magnetic fields). In the analysis of quantum chaos, in general, we encounter the eigenvalue problems of the forms $H(\tau)|n\rangle = x_n|n\rangle$ and $U(\tau)|n\rangle = \exp(-ix_n)|n\rangle$ corresponding to autonomous and driven nonautonomous systems, respectively. Here we can naturally ask a ques-

Research Reports in Physics Symmetries and Singularity Structures
Editors: M. Lakshmanan · M. Daniel

tion: if τ is regarded as "quasi-time" and both of eigenfunctions $\{|n\rangle\}$ and eigenvalues $\{x_n\}$ as dynamical variables, what will be the resultant universal dynamical system? We shall briefly give the answer to this query.

We now introduce an effective "classical" N-particle Hamiltonian with internal complex-vector space in N-dimensions for each particle [1]:

$$\mathbb{H} = \sum_{n=1}^{N} p_n^2/2 + 2^{-1} \sum_{n=1}^{N} \sum_{\substack{m=1 \\ (m \neq n)}}^{N} \langle n|L^\dagger \cdot L|m\rangle\langle m|L^\dagger \cdot L|n\rangle (x_n - x_m)^{-2}, \tag{1}$$

with $\langle n|L^\dagger \cdot L|m\rangle$ for $n \neq m$ defined as a scalar product of complex vectors $\langle n|L^\dagger$ and $L|m\rangle$. Applying Poisson brackets given by

$$\{A,B\} \equiv \sum_n \frac{\partial A}{\partial p_n}\frac{\partial B}{\partial x_n} - \frac{\partial A}{\partial x_n}\frac{\partial B}{\partial p_n} + \sum_n \frac{\partial A}{\partial \langle n|L^\dagger}\frac{\partial B}{\partial iL|n\rangle} - \frac{\partial A}{\partial iL|n\rangle}\frac{\partial B}{\partial \langle n|L^\dagger}, \tag{2}$$

one has canonical equations from (1) as

$$\frac{dx_n}{d\tau} = \{\mathbb{H}, x_n\} = \frac{\partial \mathbb{H}}{\partial p_n},$$

$$\frac{dp_n}{d\tau} = \{\mathbb{H}, p_n\} = -\frac{\partial \mathbb{H}}{\partial x_n},$$

$$\frac{d(iL|n\rangle)}{d\tau} = \{\mathbb{H}, iL|n\rangle\} = \frac{\partial \mathbb{H}}{\partial \langle n|L^\dagger},$$

$$\frac{d(\langle n|L^\dagger)}{d\tau} = \{\mathbb{H}, \langle n|L^\dagger\} = -\frac{\partial \mathbb{H}}{\partial iL|n\rangle}. \tag{3}$$

Despite an unusual appearance of the definition in (2), $\langle n|L^\dagger$ and $iL|n\rangle$ are mutually canonically conjugate--satisfying $\{\langle n|L^\dagger, iL|m\rangle\} = \delta_{n,m}$. (All theorems concerning Poisson brackets, e.g., Jacobi identity, prove to be unaltered, merely by our noting that $\{p_n\}$ and $\{\langle n|L^\dagger\}$ are on equal footing together with a similar notice for $\{x_n\}$ and $\{iL|n\rangle\}$.) Equations (1) and (3) can be identified with the quantum-mechanical eigenvalue problem [1]

$$H(\tau)|n(\tau)\rangle = (H_0 + \tau V)|n(\tau)\rangle = x_n|n(\tau)\rangle, \tag{4}$$

provided $L^\dagger$, L are defined by

$$L^\dagger \cdot L \ (\equiv \Lambda) = i^{-1}[H_0, V]. \tag{5}$$

(1) and (3) prove to describe a Calogero-Moser N-particle system in

1+1 dimensions with internal complex vector space in N-dimensions. The number of degrees of freedom of the system, (3), is N(N+1), consisting of N for positions and momenta of N particles (levels) and N^2 for internal freedom (complex eigenstates) with which each particle (level) is endowed. We can also elucidate the complete integrability of (3) [1].

Since the arguments above have been limited to autonomous systems, the next and natural question is to ask about the integrability behind nonintegrable, driven nonautonomous systems. Let us consider a quantum Hamiltonian

$$H(t) = H_0 + \tau\tilde{V}\sum_{j=-\infty}^{\infty} \delta(t - 2\pi j) ,$$

which describes any quantum bound system subjected to periodically pulsed field. H_0 and $\tilde{V}$ correspond to classically integrable part and nonintegrable perturbation, respectively, and both of them are time (t)-independent Hermitian operators ($H_0^\dagger = H_0$, $\tilde{V}^\dagger = \tilde{V}$). τ again denotes the strength of nonintegrability. For the time-dependent Schrödinger equation $i\hbar d|\Psi\rangle /dt = H(t)|\Psi\rangle$, its solution just after the j-th pulse is given by $|\Psi_j\rangle = U^j|\Psi_0\rangle$. Here U is a one-period unitary operator defined in terms of time-ordering operator T as follows:

$$U \equiv U(\tau) = T \exp\left[\int_{+0}^{2\pi+0} (-i/\hbar)H(t')dt'\right]$$

$$= \exp(-i\tau V)U_0 , \qquad (6)$$

where $V = \tilde{V}/\hbar$ and $U_0 = \exp[(-i/\hbar)2\pi H_0]$. So, the eigenvalue problem

$$U(\tau)|n(\tau)\rangle = \exp[-ix_n(\tau)]\, n(\tau)\rangle \qquad (7)$$

and its quasienergies $\{x_n(\tau)\}$ (which are discrete because of the nature of bound spectra) and quasieigenfunctions $\{|n(\tau)\rangle\}$ determine the quantum dynamics. Equation (7) proves to be identical to solving the canonical equations of motion for the "classical" Sutherland system [2].

2. Solitons as Moving Avoided Crossings

We are now standing at the stage to give a novel insight into avoided crossings. A multitude of avoided crossings in energy spectra is a clear signature of nonintegrability and chaos. Successive avoided crossings along a fictive curve are often observed, suggesting the persistence of some property of an eigenstate after several avoided crossings. We try to interpret the nature of avoided crossings from the viewpoint of nonlinear dynamics [3,4].

We have already arrived at the issue: There exists a universal "classical" dynamical system underlying the quantum chaos. The systems in (1) and (3) have been shown to be completely integrable when the number of particles is finite. Accordingly, they share the rare feature of integrability with the finite ideal gas, harmonic chain and Toda chain. Moreover, (1) and (3) reduce to the Hamiltonian for a finite ideal gas when $\langle n| L^{\dagger} \cdot L| m \rangle = 0$ for any n,m ($n \neq m$). When infinite extensions of their domains are assumed, these latter systems are known to support persistent modes of propagation. In particular, solitons can propagate in the infinite nonlinear Toda lattice. We further develop this analogy by giving here the one-soliton and the two-soliton solutions of the generalized Calogero-Moser (gCM) system.

In order to construct such solutions, we note that each solution of system (1) corresponds to a parametric family of Hamiltonians of the form $H_0 + \tau V$ in (4) and conversely. We therefore consider the following Hamiltonian operator [4]

$$H(\tau) = \begin{bmatrix} p\tau & u_1 & u_2 & u_3 & \cdots & u_N \\ u_1 & y_1 & 0 & 0 & \cdots & 0 \\ u_2 & 0 & y_2 & 0 & \cdots & 0 \\ u_3 & 0 & 0 & y_3 & \cdots & 0 \\ \cdot & \cdot & \cdot & \cdot & \cdots & \cdot \\ \cdot & \cdot & \cdot & \cdot & \cdots & \cdot \\ \cdot & \cdot & \cdot & \cdot & \cdots & \cdot \\ u_N & 0 & 0 & 0 & \cdots & y_N \end{bmatrix}, \tag{8}$$

where $y_n < y_{n+1}$. The eigenvalues of (8) are given by the roots of the equation

$$p\tau = x - \sum_{n=1}^{N} \frac{u_n^2}{x - y_n} . \tag{9}$$

This equation possesses N+1 roots $\{x_n(\tau)\}$ which generate an exact solution of the system (1) and (3). The eigenvalue problem for (8) is thus identical to solving the dynamical system (1) with initial values which are determined from the roots of (8) at $\tau = 0$. The energy spectrum is composed of N horizontal levels (i.e., independent of τ) which are crossed by one extra level with a slope given by the parameter p. In fact, each and every crossing is avoided. The repulsion between the levels is controlled by the coupling parameters $\{u_n\}$. For equal spacing between the horizontal levels, $y_n = an$, the succession of avoided crossings corresponds to a soliton propagating with a velocity p through a lattice of period a, if the horizontal levels are extended for positive and negative values of coordinate x. The exact one-soliton solution is then given by the zeros $\{x_n(\tau)\}$ of

$$p\tau = x - \frac{\pi u^2}{a} \cot \frac{\pi x}{a} . \tag{10}$$

This formula gives us the soliton profile. At a large distance from the soliton, the decrease of its profile is algebraic, $x_n(\tau) \simeq an + u^2(an - p\tau)^{-1}$. For the one-soliton solution (10), the velocity of the particles is

$$p_n(\tau) = p\left[1 + \frac{\pi^2 u^2}{a^2}\left(\sin \frac{\pi x_n(\tau)}{a}\right)^{-2}\right]^{-1} , \tag{11}$$

and the matrix elements of the interaction potential V are then

$$\langle m(\tau)|V|n(\tau)\rangle = [p_m(\tau)p_n(\tau)]^{1/2} . \tag{12}$$

The velocity (11) decreases at a large distance like

$$p_n(\tau) \simeq pu|an - p\tau|^{-2} . \tag{13}$$

Accordingly, the gCM soliton has long-range tails in contrast to the Toda soliton. The energy of the single soliton is given by

$$E_{sol} \simeq p^2u^2/2a^4 . \tag{14}$$

Solitons are known to collide with each other without undergoing a change of structure. We can exhibit this phenomenon by making a suitable generalization of the above procedure. When many solitons are created in this way, the spectrum appears irregular for any fixed value of the parameter τ. We may then associate an irregular spectrum with an ideal gas of solitons: Excitations of small and large number of solitons would correspond to the near-integrable and chaotic regions, respectively [3].

3. Statistical Mechanics of Generalized Calogero-Moser System

In previous sections we have observed that the eigenvalue spectrum is assimilated to a gas of particles moving with the pseudotime parameter τ. The collisions between particles correspond to avoided crossings of the levels. Since Hamiltonian operator in (4) works in infinite dimensional Hilbert space, we inevitably have an infinite number of energy levels. This fact leads us to consider the statistical properties of energy levels. Despite the complete integrability of the finite gCM system, statistical information is available from any finite interval in the infinite gCM system where dynamical mixing becomes possible. (Infinite dynamical systems may have ergodic, mixing, or Bernoulli properties as for the ideal gas or the harmonic solid, while no local dynamical instability exists in these systems because of the complete integrability of the corresponding finite dynamical system.) While the statistical mechanics of solitons in gCM system is an amusing subject, we shall confine in the following to the statistical mechanics of gCM system itself [5].

To study the gCM system, we need an invariant probability density defining a grand canonical ensemble. To simplify the construction, we assume that the energy spectrum extends from $-\infty$ to $+\infty$ and is invariant under x-translations with a uniform density ρ. In actual quantum systems, the level density is usually non-uniform, but for the purpose of studying the spectral fluctuations, the uniform density is desirable. The number of energy levels in any energy interval of size L large enough can thus be regarded to follow asymptotically a Poisson distribution

$$W_N^L \simeq \frac{(\rho L)^N}{N!} e^{-\rho L} . \qquad (15)$$

To construct this grand canonical ensemble, we introduce an intermediate canonical ensemble of systems with N particles in the interval $[-L/2, +L/2]$. At the end points, $-L/2$ and $+L/2$, we place hard walls where the particles undergo elastic collisions. Between the walls the motion is ruled by the classical gCM Hamiltonian (1), which is used to define a Gibbs measure

$$dM_{N,L} = \frac{1}{Z_{N,L}} e^{-\beta \mathbb{H}} \prod_{1 \le m < n \le N} dx_m \, dp_m \, d\Lambda_{mn} , \qquad (16)$$

where Λ is defined in (5). The grand canonical ensemble will then be obtained in the limit where the size L of the box increases indefinitely while keeping the density ρ constant. The resulting measure depends on two parameters: the density ρ and the inverse temperature β. The numerical value of the parameter β is determined from the variance of the velocities $\{p_n(\tau)\}$ of the energy levels:

$$\langle (\Delta p_n)^2 \rangle = \langle p_n^2 \rangle - \langle p_n \rangle^2 = 1/\beta . \qquad (17)$$

To fulfill the translational invariance of the grand canonical measure, we must now verify that a uniform density of energy levels is obtained for the canonical ensemble in the limit where $L,N \to \infty$ while N/L remains constant. This verification will eventually consolidate the foundation of random matrix theory.

The integration of the measure (16) over the variables p_m and Λ_{mn} yields the joint probability density of the energy levels:

$$f_{N,L}(x_1, \ldots, x_n) = \frac{1}{\mathbb{N}_{N,L}} \prod_{1\leq i<j\leq N} |x_i - x_j| \qquad (18)$$

for $-L/2 \leq x_i \leq +L/2$. This is just the joint probability density encountered in random matrix theory. The normalizing constant $\mathbb{N}_{N,L}$ in (18), the level density, as well as the spacing distribution will all be calculated starting from the generating functional defined by

$$I_N(u) = \int_{-1}^{1} \cdots \int_{-1}^{1} \prod_{k=1}^{N} u(x_k) \prod_{1\leq i<j\leq N} |x_i - x_j| dx_1 \ldots dx_N, \qquad (19)$$

where we set the size of the box to $L = 2$ for simplicity. The normalizing constant is then given by

$$\mathbb{N}_{N,L} = I_N(1)(L/2)^{N(N+1)/2} , \qquad (20)$$

where $I_N(1)$ is the integral (19) for the function $u(x) = 1$. Accordingly, the partition function becomes

$$Z_{N,L} = I_N(1)\ 2^{N/2}[\ (\pi/\beta)^{1/2}\ L/2\ ]^{N(N+1)/2} . \qquad (21)$$

In random matrix theory, Mehta and Gaudin [6] developed methods to reduce integrals like (19) to the Fredholm determinant of an integral kernel composed of mutually orthogonal functions. The harmonic oscillator eigenfunctions are used in the case of the Gaussian ensembles for which the joint probability density is (18) multiplied by the weight function $\exp(-\frac{1}{2}\sum_k x_k^2)$. Several other ensembles, like the Jacobi, the Laguerre, or the Legendre ensembles, have been considered by changing the weight function in order to use the corresponding orthogonal polynomials in the reduction of the integral (19) by the Mehta-Gaudin techniques. In this classification, (18) is the joint probability density of the Legendre ensembles because the weight function is 1 so that we need the Legendre polynomials to calculate (19). The procedure of the statistical mechanics of gCM system is thus reduced to the analysis of the Legendre orthogonal ensemble.

While we shall follow closely Mehta's method, the extension to the Legendre orthogonal ensemble is not straightforward. The supplementary difficulty is caused by the nonvanishing of the Legendre polynomials at the boundaries of the interval $[-1,1]$, while the harmonic oscillator eigenfunctions used for the Gaussian ensembles vanish at the boundaries of their domain of definition, namely $(-\infty,\infty)$. Choosing even values of N for simplicity, we concentrate on the calculation of the generating functional $I_N(u)$ in (19).

The domain $[-1,1]^N$ is decomposed into the N! subdomains where the variables $\{x_i\}$ are ordered. Because the integrand of (19) is a symmetric function, $I_N(u)$ is equal to N! times the integral of the same integrand but over anyone of these subdomains. For definiteness, we choose $R = \{-1 < x_1 < x_2 < \dots < x_N < 1\}$. Because the variables are now ordered, the product $\Pi|x_i - x_j|$ is equal to the Vandermonde determinant

$$\begin{vmatrix} 1 & 1 & \dots & 1 \\ x_1 & x_2 & \dots & x_N \\ \vdots & \vdots & \dots & \vdots \\ x_1^{N-1} & x_2^{N-1} & \dots & x_N^{N-1} \end{vmatrix} . \tag{22}$$

Using the properties of the Legendre polynomials that

$$P_n(x) = \frac{(2n)!}{2^n\,(n!)^2}\,x^n + O(x^{n-1}) , \tag{23}$$

the integral (19) becomes

$$I_N(u) = N!A_{N-1} \int\dots\int_R \det\Big[u(x_i)P_{j-1}(x_i)\Big]_{i,j=1,\dots,N} dx_1\dots dx_N , \tag{24}$$

with $A_{N-1} = \prod_{k=0}^{N-1} \frac{2^k(k!)^2}{(2k)!}$.

By using (24), we can calculate the normalization constant $\mathbb{N}_{N,2}$, partition function $Z_{N,L}$, uniform level density ρ and level spacing distribution p(S). In particular, the latter distribution is shown to become

$$p(S) = (\pi^2/6)S - (\pi^4/60)S^3 + (\pi^4/270)S^4 + (\pi^6/1680)S^5 + \dots . \tag{25}$$

This result agrees with the universal level spacing distribution obtained by Gaudin, Mehta and Dyson for the Gaussian and circular orthogonal ensembles [6]. Thus we understand that the statistical mechanics of gCM system can yield the major results of the random matrix theory. Furthermore, our theory can predict the distribution function of the level curvature defined by

$$K = \ddot{x}_1(\tau) = -\frac{\partial \mathbb{H}}{\partial x_1} . \tag{26}$$

Equation (26) describes τ-dependent behaviours of individual eigenvalues, thereby providing a nice indicator of quantum chaos. The probability density is defined by

$$\mathbb{P}(K) = \lim_{\substack{N,L\to\infty \\ N/L=cst}} \mathbb{P}_{N,L}(K) \tag{27}$$

with

$$\mathbb{P}_{N,L}(K) = \int \delta(K + \frac{\partial \mathbb{H}}{\partial x_1}) dM_{N,L} , \tag{28}$$

where $dM_{N,L}$ is the Gibbs measure (16). We find [5]

$$\mathbb{P}(K) = (\pi^2/2)(\rho/\beta)^2(1/|K|^3) + \dots \tag{29}$$

for large $|K|$.

In this way, nonlinear dynamics of gCM system can provide a subtle explanation of mobile avoided crossings. Statistical mechanics of gCM system, on the other hand, leads to the curvature distribution as well as the major issues of the random matrix theory, thereby suggesting a novel way to go beyond the traditional framework of random matrix theory.

My talk given here is based on a series of my joint works with collaborators, i.e., M. Lakshmanan, H. J. Mikeska, P. Gaspard and S. A. Rice. In particular, §2 and 3 are indebted to Refs. [4] and [5].

REFERENCES

[1] K. Nakamura and M. Lakshmanan, Phys. Rev. Lett. **57** (1986) 1661.
[2] K. Nakamura and H. J. Mikeska, Phys. Rev. **A35** (1987) 5294.
[3] K. Nakamura, Prog. Theor. Phys. Suppl. **98** (1989) 383.
[4] P. Gaspard, S. A. Rice and K. Nakamura, Phys. Rev. Lett. **63** (1989) 930.
[5] P. Gaspard, S. A. Rice, H. J. Mikeska and K. Nakamura, submitted to Phys. Rev. A.
[6] M. L. Mehta, **Random Matrices and Statistical Theory of Energy Levels** (Academic Press, New York, 1967).

Random Matrices and Quantum Chaos: Effects of Symmetry-Breaking on Spectral Correlations

A. Pandey

School of Physical Sciences, Jawaharlal Nehru University,
New Delhi 110067, USA

Spectral correlations in quantum chaotic systems undergo a rapid transition from one universality class to another when a symmetry of the system is gradually broken. We review the random-matrix theory for such transitions and derive the correlation functions of all orders in closed forms in the important case of time-reversal symmetry breaking.

1. Introduction

Wigner introduced the Gaussian orthogonal ensemble (GOE) of large real symmetric matrices as a model for energy level fluctuations (viz. departures from regularity) in the spectra of complex nuclei [1-5]. It is now known that the GOE serves as a model for the Hamiltonians of all autonomous complex systems whose classical analogs are fully chaotic and which are time-reversal invariant (TRI) with angular momentum good or integer-valued. Also studied are the Gaussian symplectic ensemble (GSE) of self-dual quaternion hermitian matrices and the Gaussian unitary ensemble (GUE) of (general) complex hermitian matrices, the first of these for TRI systems with angular momentum not good but half-integral and the second for time-reversal non-invariant (TRNI) systems. The three Gaussian ensembles (GE) give the three universality classes of fluctuations observed in quantum chaotic systems[4].

In all three cases the energy levels are characterized by a strong rigidity in the spectrum. The rigidity has short-range features (repulsion among neighbouring levels) as well as long-range (significant correlations among distant levels). The rigidity is moderated when there are other exact symmetries and sequences of different quantum numbers are mixed. When many sequences are mixed or when the underlying classical dynamics is completely integrable, a Poisson sequence is obtained and levels become completely uncorrelated [3,6].

There is a similar classification for the circular ensembles of unitary matrices [7,2] which serve as models for the evolution operators of systems with time-periodic Hamiltonians. The integrable case is again characterized by a Poisson sequence whereas, in the chaotic

Research Reports in Physics Symmetries and Singularity Structures
Editors: M. Lakshmanan · M. Daniel

cases, we have the three analogous universality classes: COE, CSE, and CUE of respectively symmetric, self-dual quaternion and general unitary matrices. In the latter cases the eigenangles (also called the quasi-energies) exhibit spectral rigidity.

Dyson [8] and Mehta [9] have calculated the correlation functions of all orders for the above ensembles. They have shown in particular that the Gaussian and the circular ensembles of the same symmetry class display identical fluctuation properties.

These ensembles may not however apply to systems with a partly broken symmetry. A quantum chaotic system with a small TRNI component may for example exhibit one of the following four transitions: GOE → GUE, GSE → GUE, GOE → CUE or CSE → CUE. A similar consideration for approximate quantum numbers may lead to transitions involving a few GOE's to a single GOE, Poisson → GOE, etc. In order to model these transitions, Dyson [10] introduced Brownian motion ensembles of random matrices, characterized by a fictitious time t (which measures symmetry breaking) and a fictitious temperature β^{-1} (β gives the number of distinct sites in the off-diagonal and equals 1, 4, 2 respectively for the orthogonal, symplectic and unitary cases). We have thus the Brownian ensembles of the Gaussian type $GE_{\beta}(t)$ and of the circular type $CE_{\beta}(t)$ which, for large t, approach the above equilibrium ensembles ($CE_1(\infty) \equiv$ COE, etc.). In such ensembles, the eigenvalues (constrained to move on the real line in GE and on the unit circle in CE) are like charged particles, moving under the mutual repulsion of a two-dimensional Coulomb (logarithmic) potential and executing Brownian motion. Dyson conjectured that the transition to equilibrium will be rapid, discontinuous for infinite-dimensional matrices, as a function of t.

The Brownian ensembles have been the subject of several recent studies [11-15]. The purpose of this talk is to give a brief review of the main results. Dyson's conjecture is confirmed and it is shown moreover that the transition in the correlation functions is governed, for small t and large matrix dimensionality N, by a parameter Λ which measures locally the mean square symmetry-breaking (off-diagonal) matrix element in units of local average spacing; it is zero for $t = 0$ and its large values mark the completion of the transition. For partial TRNI, correlation functions of all orders and for all Λ have been explicitly evaluated. For the other transitions, the correlation functions are given implicitly by a general hierarchic set of relations. Moreover, the correlation functions of all orders under similar initial conditions and for the same Λ and β are identical for the Gaussian and circular cases.

Section 2 will introduce the Brownian ensembles. Section 3 will give a derivation of the transition parameter and the hierarchic relations. Section 4 will give the correlation functions in the exactly-solved cases. The concluding section will discuss some applications.

2. Brownian-Motion Ensembles of Random Matrices

2.1 Equilibrium Ensembles

We first define a self-dual N-dimensional matrix B[7,16]. To take account of the three-fold classification, we write $B = \Sigma B_r e_r$ where $r = 0,1,\ldots, \beta-1$ and the e_r are the quaternion units (i.e., $e_0 = 1$, $e_r e_{r'} + e_{r'} e_r = -2\,\delta_{rr'}$ for $r, r' \neq 0$). Then, for a self-dual B, B_r is symmetric for $r = 0$ and antisymmetric for $r \neq 0$. Since $B^\dagger = \Sigma B_r^\dagger e_r$, a self-dual hermitian matrix has real B_r's. To make correspondence with quantum problems, we replace e_r by their two-dimensional matrix representatives for $\beta = 4$, but, for $\beta = 1,2$, we merely write $e_0 = 1$, $e_1 = i$. A self-dual hermitian or unitary matrix is diagonalizable by a real orthogonal matrix for $\beta = 1$, a (general) unitary matrix for $\beta = 2$, and a symplectic unitary matrix for $\beta = 4$. The eigenvalues are doubly degenerate for $\beta = 4$; usually only the distinct eigenvalues are considered but in a transition, where the degeneracy may be broken for $t \neq 0$, it must be explicitly taken into account.

The equilibrium Gaussian ensemble $GE_\beta(\infty)$ is an ensemble of self-dual hermitian matrices M in which the distinct non-zero matrix elements of M_r are distributed independently as zero-centred Gaussian variables [1-3]. The variances are v^2 for the off-diagonal matrix elements and $2v^2$ for the diagonal ones of M_0. Thus, with bar denoting ensemble averaging,

$$\overline{M}_{r:ij} = 0 , \tag{1}$$

$$\overline{M_{r:ij}M_{r':k\ell}} = \delta_{rr'}(\delta_{ik}\delta_{j\ell} \pm \delta_{i\ell}\delta_{jk})v^2 , \tag{2}$$

where in the second equation the upper sign is for $r = 0$ and the lower for $r \neq 0$. The ensemble measure is representation-independent; for example, with $\beta = 1$, all orthogonal transforms of M are contained in the GOE. This can be used to show that the joint-probability density of the eigenvalues ($x \equiv (x_1,\ldots,x_N)$) is given to within a normalization constant by

$$P(x) \propto |\Delta_N(x)|^\beta \exp(-\Sigma\, x_i^2/4v^2) , \tag{3}$$

where

$$\Delta_N(x) = \prod_{i<j} (x_i - x_j) , \quad N > 1 ,$$

$$= 1 \qquad , \quad N = 1. \tag{4}$$

The equilibrium circular ensemble $CE_\beta(\infty)$ is an ensemble of self-dual unitary matrices U with equal weight assigned to each member [7,2]. The ensemble is again representation-independent. The joint-probability density of the eigenangles ($\Theta \equiv (\vartheta_1,\ldots,\vartheta_N)$ is given by

$$P(\vartheta) \ \alpha \ |Q_N(\vartheta)|^\beta \ , \tag{5}$$

where

$$Q_N(\vartheta) = \prod_{i<j} \sin[(\vartheta_i-\vartheta_j)/2] \ , \qquad N > 1 \ ,$$

$$= 1 \qquad , \qquad N = 1. \tag{6}$$

The factors $|\Delta_N|^\beta$and $|Q_N|^\beta$ come from the Jacobian of the transformation from the matrix-element space to the eigenvariable space. The property of spectral rigidity, which arises from these factors, is thus a direct consequence of matrix diagonalization.

2.2 Non-equilibrium Gaussian Ensembles

The $GE_\beta(t)$ is an ensemble of time-dependent self-dual hermitian matrices H(t) defined for $t \geq 0$ by

$$H(t + \delta t) = H(t) + \sqrt{\delta t}\, M(t) \ , \tag{7}$$

where δt is infinitesimal and the basic Brownian step, M(t), independent for each t, is a member of $GE_\beta(\infty)$. H(0) is taken to be a diagonal hermitian matrix.

Let $x_j(t)$ be the eigenvalues of H(t). Then, from the standard perturbation theory for hermitian matrices,

$$\delta x_j(t) \equiv x_j(t + \delta t) - x_j(t)$$

$$= M_{0:jj}\sqrt{\delta t} + \sum_{k(\neq j)} \sum_{r=0}^{\beta-1} (M_{r:jk})^2 (x_j-x_k)^{-1}\delta t \ , \tag{8}$$

where in the last form, correct to first order in δt, the t-dependence of M and x are understood. For fixed x_j, we use (1,2) in H(t)-diagonal representation to perform the averages. We find [10] to the same order in δt,

$$\overline{\delta x_j} = \beta v^2 E(x_j)\delta t \ , \tag{9}$$

$$\overline{\delta x_j \delta x_k} = 2v^2 \delta_{jk}\delta t \ , \tag{10}$$

where

$$E(x_j) = \sum_{k(\neq j)} (x_j - x_k)^{-1} = \frac{\partial}{\partial x_j} \log|\Delta_N| . \tag{11}$$

A diffusion equation for the joint-probability density P(x,t) can be obtained [10] from (9,10):

$$\frac{1}{v^2} \frac{\partial P}{\partial t} = \sum_j \frac{\partial^2 P}{\partial x_j^2} - \beta \sum_j \frac{\partial}{\partial x_j}(E(x_j)P) . \tag{12}$$

Equations(9-12) also define the Brownian motion of N particles of unit charge and unit mass, moving on the real line under the influence of the mutually repulsive two-dimensional Coulomb potential ($V = -\log|\Delta_N|$). The corresponding Langevin equation is

$$\ddot{x}_j = -f\dot{x}_j + E(x_j) + W_j(t) , \tag{13}$$

where f is the friction coefficient and the W_j are white Gaussian noise ($\overline{W_j(t)} = 0$, $\overline{W_j(t)W_k(t+\tau)} = 2\beta^{-1} f \delta_{jk} \delta(\tau)$). In the overdamped case (i.e., $\ddot{x}_j \approx 0$, $\dot{x}_j$ being a slow variable) this is equivalent to (8) if we choose $\beta f v^2 = 1$ and $\beta v^2 W_j = M_{0:jj}$.

We remark here that the $GE_\beta(t)$ as defined do not have a proper stationary limit ($t \to \infty$), simply because the eigenvalues keep on moving out for large t as $t^{\frac{1}{2}}$. To have a proper limit we must rescale the matrices by a factor $(1 + \alpha t)^{-\frac{1}{2}}$ which is equivalent to including a harmonic binding term in the potential. We may then define new matrices $\mathbf{H}(\mathbf{t}) = H(t)/(1 + \alpha t)^{\frac{1}{2}}$, new eigenvalues $\mathbf{x}_j(\mathbf{t}) = x_j(t)/(1 + \alpha t)^{\frac{1}{2}}$, new time $\mathbf{t} = \alpha^{-1} \log(1 + \alpha t)$, new potential $\mathbf{V}(\mathbf{x}) = V(\mathbf{x}) + \alpha \Sigma \mathbf{x}_j^2/4\beta v^2$, new force $\mathbf{E}(\mathbf{x}_j) = -\partial \mathbf{V}/\partial \mathbf{x}_j$, and new density $\mathbf{P}(\mathbf{x},\mathbf{t}) = (1 + \alpha t)^{N/2} \times P(\mathbf{x}(1 + \alpha t)^{\frac{1}{2}}, t)$. Equations (9,10,12) remain valid under $x \to \mathbf{x}$, $t \to \mathbf{t}$, $E \to \mathbf{E}$ and $P \to \mathbf{P}$. The equilibrium density ($t \to \infty$) is given by $\mathbf{P}_{eq} \propto \exp(-\beta \mathbf{V})$ which is same as (3) for $\alpha = 1$. Since the harmonic binding does not affect the spectral correlations, we may deal with either definition.

2.3 Non-equilibrium Circular Ensembles

The $CE_\beta(t)$ is an ensemble of time-dependent self-dual unitary matrices U(t) defined by

$$U(t + \delta t) = U(t) \exp(iM(t) \sqrt{\delta t}) , \tag{14}$$

with δt and M(t) defined in Sect. 2.2 and U(0) a diagonal unitary matrix. Then, with $\exp(i\Theta_j(t))$ the eigenvalues of U(t), we have instead of (8)

$$\delta\vartheta_j(t) \equiv \vartheta_j(t+\delta t) - \vartheta_j(t)$$

$$= M_{0:jj}(\delta t)^{\frac{1}{2}} + \frac{1}{2}\sum_{k(\neq j)}\sum_{r=0}^{\beta-1}(M_{r:jk})^2 \cot[(\vartheta_j-\vartheta_k)/2]\delta t, \qquad (15)$$

and, instead of (9,10),

$$\overline{\delta\vartheta_j} = \beta v^2 E(\vartheta_j)\delta t , \qquad (16)$$

$$\overline{\delta\vartheta_j\delta\vartheta_k} = 2v^2\delta_{jk}\delta t , \qquad (17)$$

where

$$E(\vartheta_j) = \frac{1}{2}\sum_{k(\neq j)} \cot\left(\frac{\vartheta_j-\vartheta_k}{2}\right) = \frac{\partial}{\partial\vartheta_j}\log|Q_N|. \qquad (18)$$

With $x \to \vartheta$, the diffusion equation (12) gives the joint-probability density $P(\vartheta,t)$ of the eigenangles [10]. The Brownian motion of the eigenangles is again like that of an overdamped Coulomb gas but constrained to move on the unit circle with $E(\vartheta_j)$ the tangential component of the electric force felt by the charge at $\exp(i\vartheta_j)$. No rescaling is however needed to obtain the stationary limit ($P_{eq} \propto \exp(\beta\log|Q_N|)$ as in (5)).

2.4 Solution of the Diffusion Equation

We recall that the transformation $P \to P/(P_{eq})^{\frac{1}{2}}$ would give a self-adjoint form of the diffusion equation. For the Gaussian case, (12) is equivalent to

$$\frac{1}{v^2}\frac{\partial\xi}{\partial t} = -H\xi , \qquad (19)$$

where

$$H = -\sum_j \frac{\partial^2}{\partial x_j^2} + \frac{\beta(\beta-2)}{4}\sum_{j\neq k}\frac{1}{(x_j-x_k)^2} , \qquad (20)$$

and $\xi = P/|\Delta_N|^{\beta/2}$. Similarly, for the circular case, we obtain again (19) with

$$H = -\sum_j \frac{\partial^2}{\partial\vartheta_j^2} - \frac{\beta^2}{48}N(N^2-1) + \frac{\beta(\beta-2)}{16}\sum_{j\neq k}\operatorname{cosec}^2\left(\frac{\vartheta_j-\vartheta_k}{2}\right), \qquad (21)$$

and $\xi = P/|Q_N|^{\beta/2}$. With appropriate boundary conditions, the "Hamiltonian" H of (21) has well-defined eigenvalues and eigenstates in terms of which one can formally solve (12). Similarly, if we had included the harmonic binding term, the H of (20) would also have only bound states. Sutherland [17] has discussed these Hamiltonians and obtained all the bound states; no compact solution for P has emerged for general β.

However, $\beta = 2$ is a special case, since in this case the interaction terms in the Hamiltonians (20,21) drop out and then (19) becomes equivalent to a free-diffusion problem. For $GE_2(t)$ we obtain [18,13]

$$P(x,y,t) = \frac{1}{N!}\frac{\Delta_N(x)}{\Delta_N(y)} \det[g(x_i - x_j, t)]_{i,j=1,\ldots,N}, \tag{22}$$

and for $CE_2(t)$ [15]

$$P(\vartheta,\varphi,t) = \frac{1}{N!}\frac{Q_N(\vartheta)}{Q_N(\varphi)} \exp\left[\frac{tv^2 N(N^2-1)}{12}\right] \det[f(\vartheta_i - \varphi_j, t)]_{i,j=1,\ldots,N}. \tag{23}$$

Here $y \equiv (y_1,\ldots,y_N)$, are the eigenvalues of $H(0)$, $\varphi \equiv (\varphi_1,\ldots,\varphi_N)$ are the eigenangles of $U(0)$, and g, f are given by

$$g(z,t) = \frac{1}{(4\pi t v^2)^{\frac{1}{2}}} \exp\left(-\frac{z^2}{4tv^2}\right), \tag{24}$$

$$f(\psi,t) = \frac{1}{2\pi} \sum_{n=-\infty}^{\infty} \exp(-n^2 t v^2 + in\psi). \tag{25}$$

The summation index n in (25) acquires integral values if N is odd and half-integral values if N is even.

2.5 Definition of Correlation Functions

The n-level density correlation function in the GE is defined as

$$R_n(x_1,\ldots,x_n,t) = \frac{N!}{(N-n)!} \int dx_{n+1} \ldots \int dx_N \, P(x_1,\ldots,x_N,t), \tag{26}$$

where $R_N \equiv N!P$. We would be interested in the large-N limit, for which the spectrum must first be "unfolded". Using the unfolding function

$$r = \int^{x} R_1(x',t)dx', \tag{27}$$

we have the unfolded correlation function [8,3]

$$\mathbf{R}_n(r_1,\ldots,r_n,\Lambda) = \lim_{N\to\infty} \frac{R_n(x_1,\ldots,x_n,t)}{R_1(x_1,t)\ \ldots\ R_1(x_n,t)}, \tag{28}$$

where the Λ-argument in $\mathbf{R}_n$ is included in anticipation of the results ahead. One can similarly define the n-angle density correlation function in the CE ($x \to \vartheta$ in (26)). One can also define cluster functions which follow from the correlation functions [8]. All fluctuation measures, viz. measures of eigenvalue regularity used in data analysis, derive from the $\mathbf{R}_n$ [3,14,19].

3. Transition Parameter for Large N

We shall show in this section that, when $N \to \infty$, the unfolded correlation functions $\mathbf{R}_n$ for $n > 1$ undergo a discontinuous transition at $t = 0$. However for $N \to \infty$, $t \to 0$, but Λ remaining finite, a smooth transition is obtained.

We consider first [20] the large-N forms of the GE Eq. (9,10). For the level motions near X we divide the spectrum into two parts. The interaction with a cluster of P levels in the neighbourhood of X (where $N \gg P \ggg 1$) determines the fluctuating part of $\overline{\delta x_j}$ whereas the interaction with the (remaining) distant levels gives the average level motion $\overline{\overline{\delta x}}_j$ (double bar indicating an average with respect to all random variables). Writing $x \equiv \bar{x}_j$ and $\delta x \equiv \overline{\overline{\delta x_j}}$, we have then from (9)

$$\delta x = \beta v^2 \delta t\ \overline{E}(x_j) \approx \beta v^2 \delta t\ \overline{E}(x) = \beta v^2 \delta t \int_{-\infty}^{\infty} \frac{R_1(y,t)}{x-y}\, dy\ , \tag{29}$$

where a principal-value integral is used. From the conservation of probability we also have

$$R_1(x,t) = R_1(x+\delta x,\ t+\delta t)\ \frac{\partial(x+\delta x)}{\partial x}\ . \tag{30}$$

Expanding to order δt, using (29), and letting $\delta t \to 0$, we obtain

$$\frac{\partial R_1(x,t)}{\partial t} = -\ \beta v^2 \frac{\partial}{\partial x} \left\{ R_1(x,t) \int_{-\infty}^{\infty} \frac{R_1(y,t)}{x-y}\, dy \right\}, \tag{31}$$

which gives uniquely the level density $R_1(x,t)$, starting from an initial density $R_1(x,0)$. A more rigorous calculation shows that there are other terms (involving $R_2 - R_1^2$ and $\partial^2 R_1/\partial x^2$) whose neglect is justified for $N \to \infty$.

For the fluctuating part of $\overline{\delta x_j}$ we write $x_j = x + r_j D$, where $D \equiv 1/R_1(x,t)$ is the average spacing between consecutive levels and the r_j form the unfolded spectrum of the cluster. Since the contribution to $E(x_j)$ from the distant levels equals $\overline{E}(x_j)$, we have [20] (with $P \to \infty$, $N \to \infty$, $P/N \to 0$) from (9,10)

$$\overline{\delta r_j} = \beta \delta \Lambda \sum_{k(\neq j)} \frac{1}{r_j - r_k}\ , \tag{32}$$

$$\overline{\delta r_j \delta r_k} = 2\delta\Lambda\ \delta_{jk} \tag{33}$$

where $\Lambda = tv^2/D^2$ and $\delta\Lambda = \delta t v^2/D^2$. Note that the unfolding has almost completely eliminated the effect of interaction with the distant levels; its only effect is in the calculation of D and hence Λ , the latter playing the role of "new" time. Again, using the conservation of probability, we have

$$\mathbf{R}_n(r_1,\ldots,r_n,\Lambda) = \overline{\mathbf{R}_n(r_1+\delta r_1,\ldots,r_n+\delta r_n,\ \Lambda+\delta\Lambda)\frac{\partial(r_1+\delta r_1,\ldots,r_n+\delta r_n)}{\partial(r_1,\ldots,r_n)}}, \tag{34}$$

where bar denotes averaging for fixed values of the n levels $r_1,\ldots,r_n$. Expanding to order $\delta\Lambda$ and using (32,33) we obtain the BBGKY-like hierarchic relations [14,20]

$$\frac{\partial \mathbf{R}_n}{\partial \Lambda} = \sum_j \frac{\partial^2 \mathbf{R}_n}{\partial r_j^2} - \beta \sum_{j\neq k} \frac{\partial}{\partial r_j}\left(\frac{\mathbf{R}_n}{r_j - r_k}\right) - \beta \sum_j \frac{\partial}{\partial r_j} \int_{-\infty}^{\infty} \frac{\mathbf{R}_{n+1}(r_1,\ldots,r_{n+1},\Lambda)dr_{n+1}}{r_j - r_{n+1}} \tag{35}$$

where $\mathbf{R}_{n+1}$-term comes from the conditional average of δr_j over the r_k variables with $k > n$.

A similar calculation [15] for the CE Eq. (16,17) gives for R_1

$$\frac{\partial R_1(\vartheta,t)}{\partial t} = -\frac{1}{2}\beta v^2 \frac{\partial}{\partial\vartheta}\left\{R_1(\vartheta,t)\int_0^{2\pi} R_1(\varphi,t)\cot(\tfrac{\vartheta-\varphi}{2})\,d\varphi\right\}, \tag{36}$$

whereas, for the local correlations, we obtain again (32,33) and hence (35) for the $\mathbf{R}_n$. (The latter comes about because for small separations $2\cot(\frac{\vartheta-\varphi}{2}) \propto (\vartheta-\varphi)^{-1}$.) This proves that for a given β and Λ , and for similar initial conditions, the spectral correlations are identical in the Gaussian and circular ensembles.

By writing $R_1 = N\bar{\rho}$ (where then $\bar{\rho}$ is normalized to unity), we see from (31,35,36) that the transition occurs for $tv^2N \sim 1$ for the level density and for $tv^2N^2 \sim 1$ for the local correlations $\mathbf{R}_n$. There is thus [11,14] an extreme sensitivity with respect to a global symmetry breaking (measured by t) and relatively smooth behaviour with respect to a local symmetry breaking (measured by Λ). Equilibrium is reached when $\Lambda \to \infty$ for which $\partial\mathbf{R}_n/\partial\Lambda = 0$.

4. Exact Solutions of the Hierarchic Relations

The general solution of the hierarchic relations (35) is not known. However, in many cases one can obtain approximate solutions [14] and in other cases (32,33) provide efficient techniques of Monte-Carlo simulations. We list below the few cases where exact solutions have been obtained. In these cases the finite-N correlations are also known, since they derive from the direct integration of the joint-probability densities (3,5,22,23). These are

(i) GOE [9] and COE [8] ($\beta = 1$, $\Lambda = \infty$) ,

(ii) GUE [9] and CUE [8] ($\beta = 2$, $\Lambda = \infty$) ,

(iii) GSE [9] and CSE [8] ($\beta = 4$, $\Lambda = \infty$) ,

(iv) GOE → GUE [12-14] and COE → CUE [15] transitions ($\beta = 2$, arbitrary Λ, GOE or COE as initial condition), and

(v) GSE → GUE [13] and CSE → CUE [15] transitions ($\beta = 2$, arbitrary Λ , GSE or CSE as initial condition with double degeneracy explicitly taken into account).

The correlation function $\mathbf{R}_n$ can be written as the square root of the determinant of a 2n-dimensional matrix:

$$\mathbf{R}_n(r_1,\ldots,r_n) = \left\{\det[\Phi(r_i-r_j)]_{i,j=1,\ldots,n}\right\}^{\frac{1}{2}} , \qquad (37)$$

where Φ is a two-dimensional matrix of the form,

$$\Phi(r) = \begin{pmatrix} S(r) & K(r) \\ L(r) & S(r) \end{pmatrix} , \qquad (38)$$

for cases (i,ii,iv,v). Here

$$S(r) = \frac{\sin \pi r}{\pi r} , \qquad (39)$$

$$\text{(i)}\quad K(r) = -\frac{1}{\pi}\int_0^{\pi} k \sin kr \, dk , \qquad (40)$$

$$L(r) = -\frac{1}{\pi}\int_{\pi}^{\infty} \frac{\sin kr}{k} \, dk ; \qquad (41)$$

$$\text{(ii)}\quad K(r) = 0, \quad L(r) = 0 ; \qquad (42)$$

$$\text{(iv)}\quad K(r) = -\frac{1}{\pi}\int_0^{\pi} k \sin kr \exp(2\Lambda k^2) dk , \qquad (43)$$

$$L(r) = -\frac{1}{\pi}\int_{\pi}^{\infty} \frac{\sin kr}{k} \exp(-2\Lambda k^2) dk ; \qquad (44)$$

(v) $$K(r) = -\frac{1}{\pi}\int_{\pi}^{\infty} k \sin kr \exp(-2\Lambda k^2)dk\ , \tag{45}$$

$$L(r) = -\frac{1}{\pi}\int_{0}^{\pi} \frac{\sin kr}{k} \exp(2\Lambda k^2)dk\ . \tag{46}$$

K and L always appear in the form KL in (37) so that the $\Lambda\to\infty$ limits of (iv,v) are consistent with the result of (ii). For (iii) Φ is given by

$$\Phi(r) = \begin{pmatrix} S(2r) & -\frac{1}{\pi}\int_0^{\pi} k \sin 2kr\ dk \\ -\frac{1}{\pi}\int_0^{\pi} \frac{\sin 2kr}{k}\ dk & S(2r) \end{pmatrix} . \tag{47}$$

We mention here that $\mathbf{R}_2$ usually contains the most significant correlation effects. In particular long-range as well as short-range features of spectral rigidity can be deduced from this. For explicit evaluation of fluctuation measures we refer to [3,14,19].

5. Conclusion

We mention briefly the applications of random-matrix theory in problems of quantum chaos. GOE has been found to be a remarkably good model for nuclear energy level fluctuations. Not only two-level correlation effects [21] but also some higher-order correlation effects [22] have been observed. GOE also works well for atomic [23,24] and molecular spectra [25-27]. Poisson $\to$ GOE transition has been observed in atomic spectral data in connection with LS-symmetry breaking [23]. A two GOE $\to$ single GOE transition may be observed in nuclear spectra in connection with isospin-symmetry breaking [28]. GOE $\to$ GUE transition has been used for nuclear spectra to obtain bounds on the TRNI part of the nuclear interaction [14,29]. Numerical simulations of quantum billiard problems [30,31] and quantum maps [32,33] have provided extremely good confirmations of the equilibrium ensembles in chaotic systems with few degrees of freedom. Preliminary semiclassical theories have also been discussed for equilibrium [34,15] ensembles as well as one TRNI non-equilibrium [31] ensemble.

References

[1] C. E. Porter, ed., **Statistical Theories of Spectra: Fluctuations** (Academic Press, New York, 1965).

[2] M. L. Mehta, **Random Matrices** (Academic Press, New York, 1967).

[3] T. A. Brody, J. Flores, J. B. French, P. A. Mello, A. Pandey and S. S. M. Wong, Rev. Mod. Phys. **53** (1981) 385.

[4] O. Bohigas and M. J. Giannoni, in **Mathematical and Computational Methods in Nuclear Physics**, ed. J. S. Dehesa, M. G. Gomez and A. Polls, Lecture Notes in Physics 209 (Springer-Verlag, New York, 1984), pp. 1-99.
[5] O. Bohigas and H. A. Weidenmüller, Ann. Rev. Nucl. Part Sci. **38** (1988) 421.
[6] M. V. Berry and M. Tabor, Proc. R. Soc. **A356** (1977) 375.
[7] F. J. Dyson, J. Math. Phys. **3** (1962) 140.
[8] F. J. Dyson, Commun. Math. Phys. **19** (1970) 235.
[9] M. L. Mehta, Commun. Math. Phys. **20** (1971) 245.
[10] F. J. Dyson, J. Math. Phys. **3** (1962) 1191.
[11] A. Pandey, Ann. Phys. (N.Y.) **134** (1981) 110.
[12] A. Pandey and M. L. Mehta, Commun. Math. Phys. **87** (1983) 449.
[13] M. L. Mehta and A. Pandey, J. Phys. **A16** (1983) 2655.
[14] J. B. French, V. K. B. Kota, A. Pandey and S. Tomsovic, Ann. Phys. (N.Y.) **181** (1988) 198.
[15] A. Pandey and P. Shukla, to be published.
[16] M. L. Mehta, **Matrix Theory** (Hindustan Publishing Corporation, Delhi, 1989).
[17] B. Sutherland, in **Exactly Solvable Problems in Condensed Matter and Relativistic Field Theory**, ed. B. S. S. Shastry, S. S. Jha and V. Singh, Lecture Notes in Physics 242 (Springer-Verlag, Berlin, 1985), pp. 1-95.
[18] C. Itzykson and J. B. Zuber, J. Math. Phys. **21** (1980) 411.
[19] M. L. Mehta and A. Pandey, J. Phys. **A16** (1983) L601.
[20] A. Pandey, in **Quantum Chaos and Statistical Nuclear Physics**, ed. T. H. Saligman and H. Nishioka (Springer-Verlag, Berlin, 1986).
[21] R. U. Haq, A. Pandey and O. Bohigas, Phys. Rev. Lett. **48** (1982) 1086.
[22] O. Bohigas, R. U. Haq and A. Pandey, Phys. Rev. Lett. **54** (1985) 1645.
[23] N. Rosenzweig and C. E. Porter, Phys. Rev. **120** (1960) 1698.
[24] H. S. Camarda and P. D. Georgopulos, Phys. Rev. Lett. **50** (1983) 492.
[25] E. Haller, H. Köppel and L. S. Cederbaum, Chem. Phys. Lett. **101** (1983) 215.
[26] S. Mukamel, J. Sue and A. Pandey, Chem. Phys. Lett. **105** (1984) 134.
[27] Th. Zimmermann, H. Köppel, L. S. Cederbaum, G. Persch and W. Demtroder, Phys. Rev. Lett. **61** (1988) 3.
[28] G. E. Mitchell, E. G. Bilpuch, P. M. Endt and J. F. Shriner,Jr., Phys. Rev. Lett. **61** (1988) 1473.
[29] J. B. French, V. K. B. Kota, A. Pandey and S. Tomsovic, Ann. Phys. (N.Y.) **181** (1988) 235.
[30] O. Bohigas, M. J. Giannoni and C. Schmit, Phys. Rev. Lett. **52** (1984) 1.
[31] M. V. Berry and M. Robnik, J. Phys. **A19** (1986) 649.
[32] F. M. Izrailev, Phys. Rev. Lett. **56** (1986) 541.
[33] R. Scharf, B. Dietz, M. Kus, F. Haake and M. V. Berry, Europhys. Lett. **5** (1988) 383.
[34] M. V. Berry, Proc. R. Soc. **A400** (1985) 299.

Quantum Groups

C. Adimoolam

Department of Mathematics, Bharathidasan University,
Tiruchirapalli 620024, Tamil Nadu, India

This talk motivates for the study of quantum groups. Before giving the definition and examples of quantum groups, basic ideas of Hopf algebras and Poisson algebras are exposed. Quantization of Lie bialgebras is defined. At the end some open problems are mentioned.

0. Introduction

Quantization which is the process of transition from classical to quantum mechanics, involves "deforming" or "quantizing" a commutative algebra into a strictly non-commutative algebra. In classical mechanics, the states of a system are elements of a smooth manifold M, called the configuration space of the system. The phase space of the system is T*M, the cotangent bundle of the manifold M or more generally a symplectic manifold. The observables are complex valued c^{∞}-functions on the phase space. They form a commutative algebra. This algebra contains a Hamiltonian function H which governs the dynamics of the system. For example, when the phase space is T*M, then $H(p,q) = \langle p,p\rangle_q + V(q)$, where $\langle p,p\rangle_q$ is the bilinear form coming from the Riemannian structure T*M.

In the case of quantum mechanics, the states of the system are unit vectors in some Hilbert space W and the observables are the algebra of operators in W. This algebra is non-commutative containing a self-adjoint operator S corresponding to H in the classical case. The quantization should send h (Planck constant) times Poisson bracket in A to the commutator of the corresponding operators in B where A is the algebra of classical observables. We may thus consider quantization as a functor from the category of triples (M,H,A) where M is a smooth manifold, A is the algebra of complex valued c^{∞}-functions on M endowed with a Poisson structure and $H \in A$ is a Hamiltonian function, into the category of triples ($\mathbb{H}$,S,B) consisting of a Hilbert space $\mathbb{H}$, an algebra of operators on $\mathbb{H}$ containing a self-adjoint operator (the Schrödinger operator) S. The question of existence of such a functor is delicate and we will not discuss it in this paper. The

Research Reports in Physics Symmetries and Singularity Structures
Editors: M. Lakshmanan · M. Daniel

category (M,H,A) includes the category of symplectic manifolds, the symplectic form inducing the Poisson structure on $C^{\infty}(M)$. The Poisson structure is important in the mathematical study of quantization. When quantizing classical systems, one is led to quantize the Poisson symmetry groups and arrive at certain objects called quantum groups. A quantum group is not a group in the usual sense and cannot be a symmetry group of anything in the ordinary sense. Will a subgroup of a Lie group G survive as a symmetry group during the quantization of G? Thus quantum groups must play a role in symmetry breaking. To understand symmetry breaking, we need a deep study of quantum groups. For that we need an intrinsic definition of quantum groups which we lack now.

Quantum groups are obtained from solutions of Quantum Yang-Baxter equations. Sklyanin studied the classical limits of these equations which he called classical Yang-Baxter equations. From solutions of classical Yang-Baxter equations, one constructs Lie-Poisson groups and quantization of these groups enable one to construct solutions of Quantum Yang-Baxter equations. Drinfeld developed quantum groups from the point of view of Hopf algebras motivated by the study of quantum inverse scattering method. His quantum groups are the objects of a category dual to the category of Hopf algebras (not necessarily commutative nor cocommutative). Woronowicz [6] constructs quantum groups from C*-algebras. Quantum groups and their representations offer an exciting means of interpreting q-hypergeometric series. These groups have become an active area of research both for mathematicians and for physicists. There are a number of open problems in this area and I shall mention a couple of them at the end.

1. Hopf Algebras

1.1 Let K be a field and let A be an associative algebra with identity element 1 over K with multiplication $m: A \otimes A \to A$ and unit $u: K \to A$. A is called a Hopf algebra if it has three operations $\Delta: A \to A \otimes A$ (coproduct), $e: A \to K$ (co-unit) and an antipode $s: A \to A$ such that u is a coalgebra homomorphism, Δ and e are algebra homomorphisms and s is an anti-homomorphism: $s(ab) = s(b)s(a)$, $a,b \in A$ and these three maps must satisfy the following equalities:

i) for a in A, $(Id \otimes \Delta)\Delta(a) = (\Delta \otimes Id)\Delta(a)$, where $Id: A \to A$ is the identity map;

ii) $m(s \otimes Id)\Delta(a) = m(Id \otimes s)\Delta(a) = u.e(a) = e(a).1$;

iii) $(e \otimes Id)\Delta = (Id \otimes e)\Delta = Id$.

1.2 If we omit the condition of the existence of an antipode s on A, then the algebra is called a bialgebra.

1.3 Let A = Fun(G) be the algebra of functions on a group G with pointwise multiplication as multiplication operation. A is a commutative Hopf algebra but not cocommutative unless the group G is commutative. It has an antipode s defined by $sf(x) = f(x^{-1})$, $f \in A$, $x \in G$. It is easy to see that $s^2 = 1$. This is not true in the noncommutative case. In the latter case, the antipode is not unique; a second antipode s' can be defined on A for its opposite multiplication but with the same comultiplication. The two antipodes s and s' satisfy $ss' = s's = 1$.

1.4 If A is a Lie algebra over a field K, its universal enveloping algebra UA is a cocommutative Hopf algebra.

1.5 Drinfeld [3] constructs Hopf algebras by quantization which are neither commutative nor cocommutative.

1.6 By a homomorphism of Hopf algebras we mean a linear map which is both an algebra and a coalgebra homomorphism. With this definition, we see that Hopf algebras over a field (or a ring) form a category which we denote by HA.

2. Category of Quantum Groups

2.1 The category QS of quantum spaces is defined as the category dual to the category $\mathcal{A}$ of (not necessarily commutative) associative unitary algebras. If T is an object of QS and A, the corresponding algebra, we say that A is the algebra of functions on T and write T = Spec(A).

2.2 **Definition:** A quantum group is an object of QS whose associated algebra in $\mathcal{A}$ is a Hopf algebra (not necessarily commutative nor cocommutative).

2.3 Note that a quantum group is not a group object in the category of quantum spaces since in $\mathcal{A}$ the tensor product is not a coproduct in the categorical sense. Before we look at some examples of quantum groups we introduce the concept of quantization. But first we discuss briefly Poisson algebras and Lie bialgebras.

3. Poisson Algebras

3.1 **Definition:** A commutative algebra with 1, endowed with a Lie bracket $\{,\} : A \times A \to A$ which satisfies the derivation property: $\{f,gh\} = h\{f,g\} + g\{f,h\}$, $f,g,h \in A$, is called a Poisson algebra.

3.2 Products of Poisson algebras and morphisms between Poisson algebras are defined as for Lie algebras.

3.3 If M is a smooth manifold, then M is called a Poisson manifold $C^\infty(M,R)$, the algebra of real valued c^∞-functions on M, possesses a Poisson structure. Morphisms between Poisson manifolds and Poisson structure on the product of Poisson manifolds are defined in a natural manner.

3.4 A Lie-Poisson group is a Lie group with a Poisson structure such that the group multiplication $G \times G \to G$ is a Poisson morphism. Also the product structure induces a Hopf algebra structure on $C^\infty(G,R)$. Both the Hopf algebra and the Poisson algebra structures have the same multiplication. The comultiplication Δ: $C^\infty(G,R) \to C^\infty(G,R) \otimes C^\infty(G,R)$ which is an algebra map, is a homomorphism of Poisson algebras, the Poisson structure on the tensor product being given by: $\{(f \otimes g), (h \otimes k)\} = fh \otimes \{g,k\} + \{f,h\} \otimes gk$. Thus the algebra of functions on a Lie-Poisson group is a Hopf-Poisson algebra. Also, observe that the category of Hopf algebras is dual to the category of Poisson groups.

4. Lie Bialgebras

4.1 **Definition:** A Lie bialgebra over a field K is a vector space A over K endowed with both Lie algebra and Lie coalgebra structures which are compatible in the sense that the cocommutator $A \to \wedge^2 A$ is a 1-cocycle of the Lie algebra A (A acting on $\wedge^2 A$ by adjoint representation). Here by a Lie coalgebra we mean a vector space over K with a homomorphism p: $A \to \wedge^2 A$, called cocommutator, such that the homomorphism $p^*: A^* \wedge A^* \to A^*$ deduced from p is a Lie algebra structure on A^*.

4.2 Let G be a Lie-Poisson group and g its Lie algebra, g* its dual. For $a, b \in C^\infty(G)$, let $\lambda, \mu \in g^*$ such that $d_e a = \lambda$, $d_e b = \mu$, where $d_e a$ is the differential of a at the identity element e of the group G. Define $[\,,\,] = d_e\{a,b\}$. Then $[\,,\,]$ defines a Lie algebra structure on g*. Also the map $g \to \wedge^2 g$ dual to the Lie bracket $\wedge^2 g^* \to g^*$ is a 1-cocycle on g.

4.3 **Proposition:** Let G be a Lie-Poisson group with Lie algebra g. Then g is a Lie bialgebra. Further, we have the following theorem.
Theorem (Drinfeld [3]): The above correspondence $(G,P) \to (g,p)$ is an equivalence between the category of finite dimensional connected and simply connected Lie-Poisson groups and the category of finite dimensional Lie bialgebras.

4.4 **Examples.** Examples of Lie bialgebras are constructed using the (1,1) correspondence between Lie bialgebras and Manin triples

(A;B,C) where A is a Lie algebra with a nondegenerate invariant scalar product $<,>$ such that A is a direct sum of isotropic subalgebras B,C with respect to $<>$. One can easily establish this correspondence but we do not go into details here. For details, refer to [3]. Let g be a simple Lie algebra of finite dimension with an invariant scalar product $<>$. Let $A = g[u^{-1},u]$ be the Lie algebra of Laurent polynomials with coefficients from g. Write $(f,h) = \operatorname{res}_{u=0} <f(u),h(u)> du$. Let $B = g[u]$, $C = u^{-1}g[u^{-1}]$. Then $A = B \otimes C$ and (A,B,C) is a Manin triple. Thus we have a Lie bialgebra structure on B. Let $r \in g \otimes g$ correspond to the scalar product on g. Then $B \otimes B = (g \otimes g)[u,v]$. Also $A \otimes A \subset g \otimes g$. The element $s(u,v) = r/(u-v)$ is a rational function in u and v with coefficients from $g \otimes g$. The coproduct in B is defined by the formula: $f(u) = [f(u)\otimes 1 + 1\otimes f(u), s(u,v)]$. Note that B is a graded Lie algebra, graded by the degrees.

5. Quantization of Algebras

5.1 Let $0 \neq h \in \mathbb{C}$ and $\mathbb{C}[[h]]$ be the power series ring over $\mathbb{C}$. Let A_0 be a commutative algebra over $\mathbb{C}$. A deformation or quantization of A_0 is an associative algebra A_h, not necessarily commutative, over $\mathbb{C}[[h]]$, topologically free such that $A_h/hA_h = A_0$. If A_0 is a Lie bialgebra, by a deformation of A_0 we mean a deformation of its universal enveloping algebra in the above sense.

5.2 If A_h is a quantization of A_0, then A_0 can be given the structure of a Poisson algebra by defining, for $a_0,b_0 \in A_0$, $\{a_0,b_0\} = h^{-1}[a,b]$ mod h, where $a,b \in A_h$ with images a_0,b_0 respectively under the canonical map. This definition depends only on a_0, b_0. It is easy to check that $\{,\}$ is a Poisson bracket on A_0. If (A,P) is a Hopf-Poisson algebra (i.e., a commutative Hopf algebra endowed with a Poisson bracket such that the comultiplication preserves the Poisson structure), its quantization is a Hopf algebra A_h over $\mathbb{C}[[h]]$, topologically free such that $A_h/hA_h = A$, inducing on A the Poisson structure P. The algebra (A,P) is called the classical limit of A_h. There is a dual notion of Hopf co-Poisson algebra.

5.3 **Remark:** The connection between the quantization defined above and Gerstenhaber's deformation theory of algebras and the obstruction to quantizing a Lie algebra will be explained in a forthcoming paper. It is not known whether every Lie bialgebra can be quantized.

6. Yang-Baxter Equations

6.1 In this section, we shall study quantum Yang-Baxter equations (QYBE) and their classical limits-classical Yang-Baxter equations (CYBE). QYBE play a profound role in statistical mechanics, integrable systems, S-matrix theory and conformal field theories. They play in quantum groups (or q-deformation of Lie algebras) a role analogous to the Jacobi identity in Lie algebras. It will be fruitful to study Yang-Baxter categories which are tensor categories in which Yang-Baxter equations hold.

6.2 Let V be an n-dimensional vector space over $\mathbb{C}$. Let R(x) be an endomorphism of $V \otimes V$, depending on a parameter x called the spectral parameter which is the rapidity difference between two colliding particles. By choosing a basis $e_1, e_2, \ldots, e_n$, R(x) is represented by an $n^2 x n^2$-matrix $R^{ik}_{jl}(x)$. In two dimensional lattice theory, $R^{ik}_{jl}(x)$ may be viewed as the Boltzmann weight attached to each vertex of the lattice. Define $R^{12} = R \otimes Id_3 \in End(V \otimes V \otimes V)$, i.e., R^{12} acts as R on the first two components and as identity on the third component. Similarly, R^{13} and R^{23} are defined. Then the quantum Yang-Baxter equations are written as $R^{12}(x_1-x_2)R^{13}(x_1-x_3)R^{23}(x_2-x_3) = R^{23}(x_2-x_3)R^{13}(x_1-x_3)R^{12}(x_1-x_2) R^{12}(x)R^{21}(-x) = 1$.

6.3 It is an interesting problem to find the complete set of solutions of QYBE. Does the solution set have a "nice structure"?

6.4 Suppose QYBE have solutions of the type $1+r(x)h + \ldots$. Then substituting in QYBE and making h tend to 0, we obtain classical Yang-Baxter equations(CYBE):

$$[r_{12}(x_1-x_2), r_{13}(x_1-x_3)] + [r_{12}(x_1-x_2), r_{23}(x_2-x_3)] +$$

$$[r_{13}(x_1-x_3), r_{23}(x_2-x_3)] = 0$$

$$r_{12}(x) + r_{21}(-x) = 0.$$

6.5 Note that since the CYBE are expressed in terms of brackets, the equations make sense for a $g \otimes g$-valued function r(x) for an abstract Lie algebra g. Suppose g is finite dimensional and let e_i be a basis of g. Suppose also that g is simple. Then $r(x) \in g \otimes g$ can be written as $r(x) = r_{ij}(x)$. Belavin and Drinfeld [2] consider meromorphic solutions r(x) defined on a circle in $\mathbb{C}$ with centre at the origin.

6.6 **Theorem** (Belavin-Drinfeld): One such solution extends meromorphically to the entire complex plane and all its poles are simple.

The set of poles form a discrete subgroup of $\mathbb{C}$. If the rank of the discrete subgroup L is 2, then the solution is an elliptic function. If the rank of L is one, then r(x) is equivalent to a trigonometric solution and if the rank of L is zero, then r(x) is equivalent to a rational solution. (Note: Here solution means always nondegenerate solution, i.e., the determinant of $r_{ij}(x)$ is not identically zero.)

7. Examples of Quantum Groups

7.1 Let $R \in \mathrm{End}(\mathbb{C}^n \otimes \mathbb{C}^n)$ be a quantum Yang-Baxter operator, i.e., R satisfies QYBE. Let A be the free algebra generated by 1 and n^2 elements t_{ij}, $1 \le i, j \le n$ over $\mathbb{C}$. Set $T = (t_{ij})$, $T_1 = T \otimes I$, $T_2 = I \otimes T$. Suppose T is invertible and satisfies the following equation: $RT_1T_2 = T_2T_1R$. Let I_R be the two-sided ideal generated by the relations $RT_1T_2 = T_2T_1R$. The quantum space corresponding to the quotient algebra A/I_R is the space of matrices of rank n.

7.2 Let us return to the Lie bialgebra of example 4.4. One can show that the algebra B has a unique graded deformation B_h with deg(h) = 1. In [3], this algebra has been described by generators and relations. The degree zero part $B_{h,0}$ of B_h is the universal enveloping algebra of g. B_h is generated by $B_{h,0}$ and a $\mathbb{C}$-vector space in degree one isomorphic to g. By putting h = 1 in the relations defining B_h over $\mathbb{C}[[h]]$, we obtain a quantum group called the Yangian of g, denoted by Y(g).

7.3 Let g be a simple Lie algebra as above. Let r = rank of g. Let G be the Lie group corresponding to g. Let $\alpha_1, \alpha_2, \ldots, \alpha_r$ be the simple roots of g. $T = (t_{ij})$ is the Cartan matrix of g with respect to the invariant scalar product of g. The quantum enveloping algebra $U_h g$, $h \in \mathbb{C}$ of the Lie algebra g, introduced by Drinfeld [3] and Jimbo [4] is the algebra over $\mathbb{C}[[h]]$, generated by H_i, $X_i^\pm$, $i = 1, \ldots, r$ and the relations $[H_i, H_j] = 0$, $[H_i, X_j^\pm] = \pm(\alpha_i, \alpha_j)X_j^\pm$

$$[X_i^+, X_j^-] = \frac{e^{hH_i} - e^{-hH_i}}{e^h - e^{-h}} \delta_{ij}, \quad i,j = 1, \ldots, r$$

and

$$\sum_{k=0}^{m} (-1)^k \binom{m}{k}_{q_i} q_i^{-k(m-k)/2} (X_i^\pm)^k X_j^\pm (X_i^\pm)^{m-k} = 0 \text{ if } i \neq j$$

where $m = 1 - t_{ij}$, $q_i = \exp[h(\alpha_i, \alpha_i)]$ and $\binom{m}{k}_q$ is the q-binomial coefficient.

In the limit when $h \to 0$, we get the universal enveloping algebra of g. We call Spec(U_hg) the quantum group obtained by quantizing the Lie group G. That U_hg is indeed a Hopf algebra is left as an exercise.

7.4 Drinfeld studied in [3] the relation between Y(sl(n)), U_h(sl(n)) and Hecke algebras.

7.5 On the other hand, Woronowicz [6] constructs compact quantum groups starting from a C*-algebra. Let A be a C*-algebra with unity. $T = (t_{ij})$ an n×n-matrix with entries from A. Then a compact quantum group is said to be given by the pair (A,T) if, for the involutive subalgebra B generated by the t_{ij}, we have:

1. B is dense in A ;
2. there is a C*-algebra homomorphism $\Delta : A \to A \otimes A$ such that $\Delta(t_{ij}) = \Sigma_j t_{ij} \otimes t_{jk}$;
3. there is an antipode $s: B \to B$ such that $s(s(a^*)^*) = a, \forall\, a \in B$.

We do not discuss in this paper the connection between this definition and the one defined by Drinfeld.

8. Remarks and Open Problems

8.1 There are many important topics relating quantum groups which I have not touched here. For example, I have not said anything at all about quantum linear algebra and quantum algebraic groups and the connection between quantum groups and braid group representation and invariants of links.

8.2 What role will the quantum groups play in algebraic geometry, in particular, arithmetic algebraic geometry?

8.3 Classify the representations of quantum groups and construct models for the representations of these groups. Representations of U_hSU(n) run parallel to those of SU(n) [5,6]. Explain the quantum analogue of the connection between special functions and compact semi-simple Lie groups.

8.4 What is the connection between the representations of quantum groups and conformal field theories?

8.5 Can we develope a coherent quantum theory with quantum groups playing the role of symmetry groups?

These are only a few of the many problems which arise in the exciting study of quantum groups.

Acknowledgment

This paper is based on my talk at the workshop on "Symmetries and Singularity Structure Aspects of Nonlinear Dynamical Systems" held during November-December 1989 at Bharathidasan University, Tiruchirapalli. A detailed account of the topics and problems mentioned in this paper will be published in the future. Finally, I thank Prof. M. Lakshmanan, Department of Physics, Bharathidasan University for giving me the opportunity to expose the ideas of quantum groups.

References

[1] C. Adimoolam, Quantum groups and quantum geometry, in preparation.
[2] A. A. Belavin and V. G. Drinfeld, Triangle equations and simple Lie algebras, Math. Phys. Rev. **4** (1984).
[3] V. G. Drinfeld, Quantum groups, Proc. Int. Congr. Math. Berkeley, 1986.
[4] M. Jimbo, A q-analogue of U(gl(N+1)), Hecke algebras and the Yang-Baxter equations, Lett. Math. Phys. **11** (1986) 247.
[5] T. H. Koornwinder, Representations of the twisted SU(2) quantum group and some q-hypergeometric orthogonal polynomials, preprint.
[6] S. Woronowicz, Compact matrix pseudogroups, Comm. Math. Phys. **111** (1987) 287.

Integrable Quantum Spin Chains and Some Problems Related to Integrable Systems

A. Kundu and B. Basumallick

Theoretical Nuclear Physcis Division, Saha Institute of Nuclear Physics, 92 APC Road, Calcutta 700009, India

The correspondence between integrable systems and conformally invariant models is explored through the example of nonlinear σ-models to find some intriguing relation between integrability, boundary condition and conformal symmetry of the model. In the second part, an integrable generalisation of the XYZ model with higher spins is presented and some novel symmetry of the related R-matrix is found out leading to many anisotropic integrable models.

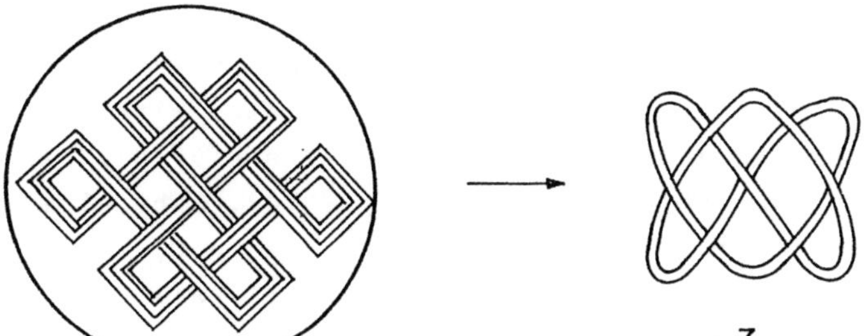

7_4

This knot is one of the eight glorious emblems of Tibetan Buddhism. It symbolises the interdependence of all things in this phenomenal world. This is equivalent to the famous 7_4 knot and the knots also do not exist without reference to its embedding in space [1].

Recent years witness appearance of extremely interesting interrelationship between seemingly different subjects like quantum and statistical integrable systems, integrable system and quantum groups, Yang-Baxter equation, braid groups and fractional statistics, etc. Since a thorough discussion of all these subjects is beyond the scope of this lecture we would concentrate only on some intriguing aspects of the relationship between integrable and conformal field theories. This would be dealt with in the first part of the lecture, where we compare some essential properties of these systems and present our result related to the SU(2) σ-model. In the second part we introduce an integrable higher spin generalisation of the XYZ chain and find some novel symmetries of the Yang-Baxter equation leading to various anisotropic integrable models.

Research Reports in Physics — Symmetries and Singularity Structures
Editors: M. Lakshmanan · M. Daniel © Springer-Verlag Berlin, Heidelberg 1990

1. Integrable and Conformal Limits of SU(2) Nonlinear σ-Model

The fact that both integrability and conformal invariance are attributed to two-dimensional field theories, is in a sense indicative of some relationship between them. In spite of many serious work and interesting conjectures [2], this link is yet to be understood fully. Before going further let us explain first the notion of integrability, which would be understood in the Liouville sense generalised for the infinite dimensional system. An N-dimensional system is said to be integrable if there exist action-angle variables $\{I_\iota, \varphi_\iota\}$ such that the resultant Hamiltonian becomes a function of action variables only, leading to the equations $\dot{I}_i = 0$ and $\dot{\varphi}_i = \frac{\partial H}{\partial I_i} = \omega_i$. Consequently, the system is trivially integrated giving $I_i = I_{0i}$, $\varphi_i = \omega_i t + \varphi_{0i}$, with I_{0i} as the N number of independent conserved quantities. Therefore generalisation of the notion of integrability to field theoretical models demands the existence of infinite number of conserved quantities in involution, which should include all the observables like Hamiltonian, momentum, etc. and must be expressible through the action variables only. It is also well known that integrable systems are associated with a linear spectral problem and the related matrix elements $a(\lambda)$ and $b(\lambda)$ play the role of 'action' and 'angle' variables, respectively [3,4]. The element $a(\lambda)$ exhibits the following important properties: (i) $\dot{a}(\lambda) = 0$, (ii) $a(\lambda)$ is analytic with no poles at $\mathrm{Im}\lambda \geq 0$, which naturally allows Laurant series expansion of the type

$$\log a(\lambda) = \begin{cases} \sum_{n=1}^{\infty} C_n \lambda^{-n}, & \text{at } \lambda \to \infty \\ \sum_{n=1}^{\infty} D_n \lambda^{n}, & \text{at } \lambda \to 0 \end{cases} \tag{1.1}$$

(iii) the 'action' variables $a(\lambda)$ are in involution: $\{a(\lambda), a(\mu)\} = 0$. Note that these properties are most crucial for the integrability of the system. For example, (i) and (ii) yield that C_n, D_n do not depend on time and they constitute the infinite set of conserved quantities required for the integrability. Finally, (iii) guarantees that these conserved charges are in involution. Another natural requirement is that all the conserved quantities including Hamiltonian (H), momentum (P), etc. of the system must be expressed through the conserved set C_n, D_n acting as 'action' variables. We observe also that H and P have canonical dimension 1 in relativistic models, which should therefore must be constructed from C_1 and D_1. On the other hand, since $\log a(\lambda)$ is dimensionless and so is λ for the relativistic models [5],

C_n and D_n must also be dimensionless. But in that case how they can generate dimensional physical quantities like H, P, etc.? The remedy from this apparent paradox is the following. In integrable models there exists a dimensional 'mass' parameter m and, in fact, it enters the conserved charges as $C_n = C_n'/(m^{n+1})$ and $D_n = D_n'/(m^{n+1})$. consequently, C_n, D_n acquire canonical dimensions and may therefore generate dimensional observables like H, P, etc. In the light of the above discussions we may compare now the properties of the integrable systems with those of the conformal theories.

1. Integrable systems (IS) are defined basically at finite interval with periodic boundary condition and subsequently, the period may be taken to infinity to get the model at $L \to \infty$, the conformal symmetry (CS) on the other hand is exhibited only at ∞-interval and breaks down at $L < \infty$.
2. IS requires analyticity of $a(\lambda)$ in spectral parameter λ, while CS needs the same for $T(z)$ in the configuration space.
3. The generators of conserved charges in IS are in involution, while the conformal generators satisfy Virasoro algebra and are not independent.
4. Presence of dimensional 'mass' parameter is essential for integrability; CS is spoiled by 'mass' parameter.
5. Integrability is an off-critical phenomena, while CS might appear only at criticality.

The above comparison shows that both these properties cannot coexist under the same condition and as the conformal limit is approached most of the features of integrability is destroyed, though some properties for certain integrable systems might have smooth transition at criticality. To have a closer look of this relationship let us consider SU(2) nonlinear σ-model, which is known to possess both of the properties under discussion. The model may be given through the equation of motion

$$\partial_t I_\pm(x) = \pm\partial_x I_\pm \pm [I_+, I_-] \; ; \qquad I_\pm = g^{-1}\partial_\pm g$$

in the light-cone coordinates $x_\pm = (t \pm x)/2$. The corresponding Hamiltonian and momentum of the system may be given by

$$H = \int tr(I_+^2 + I_-^2)dx \; , \quad P = \int tr(I_+^2 - I_-^2)dx \; . \tag{1.2}$$

The model allows linear Lax pair operators and is believed to be integrable [6]. Nevertheless we aim at checking its integrability property in the sense discussed above for the natural boundary condition: $g(x) \to 1$. Starting from the Lax operators and through the Riccati equation we may generate in a standard way [5] the conserved quantities as the expansion coefficients of

$$\log a(\lambda) = \sum_{n=-1} C_n/\lambda^{-n}, \text{or} = \sum_{n=-1} D_n\lambda^n , \tag{1.3}$$

where $C_n = \int c_n\, dx$ and $D_n = \int d_n\, dx$ with

$$c_{-1} = (trI_+^2)^{1/2}, \quad d_{-1} = (trI_-^2)^{1/2} ,$$

$$c_0 = c_{-1} - f_c , \quad d_0 = d_{-1} - f_d , \tag{1.4}$$

where f_c and f_d are some complicated functions [5]. It can be shown that these conserved charges are in involution and thus obey the primary requirement of integrability. However, in spite of this favourable condition, it is surprising to note that neither H nor P is expressible through these action variables C_n and D_n, which is evident from the comparison of (1.2) and (1.4). This unusual feature is possibly explained by the fact that the analyticity of $a(\lambda)$ is spoiled here by the appearance of a pole in (1.3). Another reason of this subtlety is the absence of any 'mass' parameter in the system. From these facts, we can conclude that this conformally invariant model with trivial BC is not an integrable system. Therefore, to find a possible integrable limit we consider a nontrivial BC given by $g \to g_0(mx)$ compatible with $I_{\pm} \xrightarrow[|x|\to\infty]{} m_\pm \sigma_3$ by introducing new dimensional parameters $m_\pm$. Moreover we impose the constraint $trI_\pm^2 = m_\pm^2$, $m_\pm \neq 0$ by fixing a particular conformal gauge. It is amazing to note that the situation is completely changed now with the expansion of $\log a(\lambda)$ given as

$$\log a(\lambda) = \sum_{n=0} \tilde{C}_n/\lambda^{-n}, \text{or} = \sum_{n=0} \tilde{D}_n\lambda^n , \tag{1.5}$$

i.e., the required analyticity is restored and more than that the conserved charges are now given by

$$\tilde{C}_n = \tilde{C}_n'/m_+^{n+1} \quad , \quad \tilde{D}_n = \tilde{D}_n'/m_-^{n+1} \tag{1.6}$$

with the Hamiltonian, momentum of the new system [5,7] expressible through the 'action' variables as $H = \tilde{C}_0' - \tilde{D}_0'$ and $P = \tilde{C}_0' + \tilde{D}_0'$, etc. Thus all the Liouville criteria are fulfilled and we conclude that the system is integrable with this particular BC. On the other hand, it may be shown [8] that the conformal symmetry of the model is not preserved in this case including its Lorentz invariance. Our result is also consistent with Zamolodchikov conjecture [2] that for getting an integrable system one has to break the conformal symmetry of the model by a relevant perturbation.

2. Integrable Anisotropic Spin Chains and Related R-Matrix

Heisenberg spin chains are known to be integrable for spin half values. Recently integrable generalisation of these models with arbitrary spins has been proposed [9-11]. We aim here to formulate an alternative generalisation of the anisotropic XYZ case with higher spins through direct matrix operator formulation. We also investigate the symmetries of the related quantum R-matrix and with their help generate new integrable models. Our construction of R-matrix through higher representation of SU(2) generators is different from that of the fusion method [11] as well as from the formulation of the matrix through some prescribed conditions [12]. This on the one hand helps us to extract the symmetries of the R-matrix as a consequence and on the other hand allows us to detect the rule for nonvanishing of its elements. At the same time it yields directly the Hamiltonian of the system through spin operators [13]. The proposed R-matrix through higher spin generators S^{α} , α = 1,2,3 is taken in the form

$$R^{s}_{xyz}(\lambda) = \sum_{k=1}^{2s} \sum_{\{\alpha_k\}} f_{\alpha_1 \alpha_2 \cdots \alpha_k}(\lambda)\, \Omega^{\alpha_1} \Omega^{\alpha_2} \cdots \Omega^{\alpha_k} + f_0, \qquad (2.1)$$

where $\Omega^{\alpha_k} = S^{\alpha_k} \otimes S^{\alpha_k}$, with $\{\alpha_k\}$ denoting permutation of all three spin matrix components and the unknown functions $f_{\{\alpha\}}$ with total symmetry in their indices are $N^{s}_{xyz} = (2s+1)(2s+2)(2s+3)/6$ in number and they should be determined from the Yang-Baxter equation. It can be shown that the construction (2.1) fixes the symmetries of the R-matrix, e.g., C, P and T symmetries and finds the nontriviality condition for its matrix elements R^{kl}_{ij} as $[(i+j) - (k+l)] = 2N$, $N = 0,1,2, \ldots,2s$; i.e., the difference between 'incoming' and 'outgoing' spins may be given only by an even number limited by a maximum value 4s. It should be mentioned that the advantage of our construction is that, we get these symmetries as a natural consequence, while they are generally needed to be imposed *apriori* [12]. The corresponding Hamiltonian of the model may be given through

$$H = -\partial_{\lambda} [T(\lambda)]_{\lambda=0} - CN1_N ,$$

where $T_N(\lambda) = tr(\prod_n \mathbb{L}_n(\lambda))$ with $\mathbb{L}_n(\lambda)$ constructed out of R-matrix in the standard way. In our construction one gets the Hamiltonian of the system directly through the higher spin operators. Using our ansatz we are able to construct explicitly the integrable spin-one XYZ chain with anisotropic Hamiltonian structure having interesting crystal field terms [13].

Another important fact about these R-matrices is that new matrices may be generated from the original ones by exploiting the symmetries of the Yang-Baxter equation [14]. However, these symmetries are known only for charge conserved models while the XYZ model considered here does not respect such a conservation law. Therefore, we have to search for some generalized symmetries. Fortunately we find that the transformation

$$R_{ij}^{kl} \longrightarrow \exp[\{[f(i) + f(j)] - [f(k) + f(l)]\}\omega]R_{ij}^{kl} \tag{2.2}$$

with arbitrary function f also answers to the symmetries of the Yang-Baxter equation and amazingly this is a model-independent symmetry and consequently applicable naturally to the XYZ model. Therefore, new R-matrices may be generated from the original one given by (2.1), which in turn may yield new integrable models with new Hamiltonian operators having more assymetric terms. This procedure is demonstrated for the particular case of s = 1 in our recent work [13]. For s = 1/2 the known 8V(1) type model [15] may be obtained easily following this approach. Similarly such symmetries for the XXZ and XXX models with arbitrary spin can also be exploited for finding new integrable models as well as for generating known models like 6V(1) [15].

It would also be an interesting problem to explore the possibility of applying such transformation of the R-matrix for finding some deformation (if any) of the respective quantum group associated with such integrable spin models.

REFERENCES

[1] D. Rolfsen, **Knots and Links** (Publish or Perish, Berkeley, 1976).
[2] A. B. Zamolodchikov, JETP Lett. **46** (1987) 160.
[3] S. Novikov, V. Manakov, L. Pitaevskii and V. Zakharov, **Theory of Solitons** (Plenum, New York, 1984).
[4] L. D. Faddeev, **Les Houches Lectures**, ed. J. Zuber and S. Stora (North-Holland, Amsterdam, 1984), p. 561.
[5] S. Ghosh and A. Kundu, Phys. Rev. Lett. **63** (1989) 1207.
[6] K. Pohlmayer, Comm. Math. Phys. **46** (1976) 207.
[7] L. D. Faddeev, proc. 7th Int. Conf. on the Problems of QFT, Alushta (Dubna, USSR, 1984), p. 34.

[8] S. Ghosh, SINP Preprint No. SINP/TNP-89-25, 1989.
[9] P. Kulish, N. Reshetikhin and E. K. Sklyanin, Lett. Math. Phys. **5** (1981) 393.
[10] A. Kirillov and N. Reshetikhin, J. Phys. **A20** (1987) 1565.
[11] E. Date, M. Jimbo, M. Miwa and M. Okado, Lett. Math. Phys. **12** (1986) 209.
[12] V. A. Fateev, Sov. J. Nucl. Phys. **33(5)** (1981) 761.
[13] B. Basumallick, A. Kundu, SINP preprint No. SINP/TNP/90-2, 1990.
[14] M. Wadati, T. Deguchi and Y. Akutsu, Phys. Reports **180** (1989) 247.
[15] K. Sogo, M. Uchinomi, Y. Akutsu and M. Wadati, Prog. Theor. Phys. **68** (1982) 508.

On the Quantum Inverse Problem for a New Type of Nonlinear Schrödinger Equation for Alfven Waves in Plasma

S. Sen[1] *and A. Roy Chowdhury*[2]

[1]International Centre for Theoretical Physics, Trieste, Italy
[2]Department of Physics, High Energy Physics Division, Jadavpur University, Calcutta 700032, India

The nonlinear Alfven waves are governed by the vector derivative nonlinear Schrödinger (VDNLS) equation, which for parallel or quasi-parallel propagation reduces to the derivative nonlinear Schrödinger (DNLS) equation for the circulary polarized waves [1].

We have formulated the Quantum Inverse problem for this new type of nonlinear Schrödinger equation written as

$$\psi_{1t} = i\psi_{1xx} - 2c\psi_1\psi_2\psi_{1x}$$

$$\psi_{2t} = -i\psi_{2xx} - 2c\psi_1\psi_2\psi_{2x} .$$

This set of evolution equations have many properties similar to the usual NLS problem but the structure of classical and quantum R-matrix are distinctly different. The commutation rules for the scattering data are obtained and the algebraic Bethe Ansatz is formulated to derive the eigenvalue equation for the energy of the excited states. An important feature of our problem is that the symplectic structure associated is very similar to the DNLS equation and it was impossible to formulate quantum inverse scattering method (QISM) for DNLS.

1. Introduction

The quantization of nonlinear systems is one of the most important programmes in the study of solitons. Apart from the semiclassical approach of Gutzwiller [2], the most important formulation was that of Faddeev [3]. The basic ingredient of the formulation is the classical $r(\lambda,u)$ and quantum $R(\lambda,u)$ matrices. Theories with some particular form of r and R have been classified by Tarasov [4] and Sklyanin [5]. Here we have tried to formulate the quantum inverse problem for a new integrable nonlinear system, similar to the NLS problem. Usually the approach of Faddeev and Sklyanin is not applicable to a case of non-ultralocal system. Due to this reason it was not possible to construct a QISM framework for the original DNLS problem. Here in this communication we deal with an equation similar to the DNLS problem but with a symplectic form which

Research Reports in Physics Symmetries and Singularity Structures
Editors: M. Lakshmanan · M. Daniel

is ultralocal. So it is possible to apply the technique of QISM. Such an equation has been used in the propagation of Alfven waves in a plasma. In the following sections we give the details of our result starting with the construction of the quantum R-matrix, the commutation rules for the scattering data, the modified form of the Yang-Baxter equation and the eigenvalue problem for the excitation.

2. Formulation

The equations under consideration can be written as [6]

$$\psi_{1t} = i\psi_{1xx} - 2c\psi_1\psi_2\psi_{1x}$$

$$\psi_{2t} = i\psi_{2xx} - 2c\psi_1\psi_2\psi_{2x} \tag{1}$$

Equation (1) is an integrable system both in the sense of Painlevé analysis and in having a Lax pair. The Lax pair has a similar form regarding its dependence on the field variables. But with respect to the eigenvalue parameter λ, it resembles the Kaup-Newell problem. The space part of the Lax pair for (1) can be written as

$$\left[\frac{\partial}{\partial x} + \begin{pmatrix} -\frac{1}{2}(c\psi_1\psi_2 + \lambda^2) & -\lambda\psi_1\sqrt{c} \\ -\lambda\psi_2\sqrt{c} & \frac{i}{2}(\lambda^2 + \psi_1\psi_2\ c) \end{pmatrix}\right]\psi = 0\ . \tag{2}$$

At present there exist two approaches to the quantization:

a) The differential equation approach: this method has been critized severely by Gutkin [7].

b) The space discretization approach: this does not suffer from difficulties of (a) but is still considered to be an approximate one unless the limit $N\to\infty$ is properly taken, where N is the number of subdivisions in the interval (0,L) of the real axis in subintervals of length Δ. We pursue here the second line of working.

By converting the linear problem into a Riccati system, we can generate an infinite number of conservation laws, C_n. From these it is easy to pick up the Hamiltonian given as

$$H = \int_{x_0}^{x_0+L} C_2\ dx = \frac{1}{2}\int_{x_0}^{x_0+L} [i(\psi_1\psi_{2xx} - \psi_{1x}\psi_{2x})$$

$$- \psi_1\psi_2\ (\psi_1\psi_{2x} - \psi_{1x}\psi_2)]dx \tag{3}$$

from which (1) can be generated via

$$\frac{\partial\psi_1}{\partial t} = \{H,\psi_1\}_{\mathbb{L}}$$

$$\frac{\partial\psi_2}{\partial t} = \{H,\psi_2\}_{\mathbb{L}} \tag{4}$$

where $\mathbb{L}$ is the symplectic operator

$$\begin{pmatrix} 0 & -i \\ i & 0 \end{pmatrix} \tag{5}$$

yielding the Poisson bracket

$$\{f,g\}_{\mathbb{L}} = i\int_{x_0}^{x_0+L}\left(\frac{\delta f}{\delta\psi_2}\frac{\delta g}{\delta\psi_1} - \frac{\delta f}{\delta\psi_1}\frac{\delta g}{\delta\psi_2}\right)dx \ . \tag{6}$$

Thus we proceed with the discretization of the interval (x_0,x_0+L) and rewrite the Lax equation (2) in the form

$$\psi_{n+1} = L_n\psi_n \tag{7}$$

where L_n is defined via

$$L_n \approx 1 + \Delta\int_{x_0}^{x_0+L} L\ dx \tag{8}$$

and

$$\psi_{1n} = \int_{x_0}^{x_0+L} \psi_1\ dx$$

$$\psi_{2n} = \int_{x_0}^{x_0+L} \psi_2\ dx \ . \tag{9}$$

The Poisson bracket is defined through

$$[\psi_{1n},\ \psi_{2n}] = -\frac{i}{\Delta} \tag{10}$$

and

$$L_n = \begin{pmatrix} 1 - \frac{i\Delta}{2}(c\psi_{1n}\psi_{2n} + \lambda^2) & -\Delta\lambda\psi_{1n}\sqrt{c} \\ -\Delta\lambda\psi_{2n}\sqrt{c} & 1 + \frac{i\Delta}{2}(\psi_{1n}\psi_{2n}\ c + \lambda^2) \end{pmatrix} \tag{11}$$

2a The Classical r-matrix

The classical r-matrix is defined following Faddeev via the equation

$$\{L(\lambda,x) \otimes L(u,y)\} = [r(\lambda,u),\ L(\lambda,x)\otimes 1 + 1\ \ L(u,y)] \tag{12}$$

Computing the 16 Poisson brackets on the LHS we can easily solve for $r(\lambda,u)$, and

$$r(\lambda,u)=\begin{bmatrix} -\frac{(\lambda^2+u^2)}{2(\lambda^2-u^2)} & 0 & 0 & 0 \\ 0 & 0 & -\frac{\lambda u}{\lambda^2-u^2} & 0 \\ 0 & -\frac{\lambda u}{\lambda^2-u^2} & 0 & 0 \\ 0 & 0 & 0 & -\frac{(\lambda^2+u^2)}{2(\lambda^2-u^2)} \end{bmatrix} \tag{13}$$

This form of r-matrix is distinctly different from that of the NLS model, having poles at two positions $\lambda = \pm u$, but the symmetry remains the same.

2b Quantum R-matrix

For the quantum mechanical case we start with the discrete form of L, i.e., L_n and observe that πL_n can be interpreted as the transition matrix. We now evaluate [8]

$$R(\lambda,u)[L_n(\lambda) \otimes L_n(u)] = [L_n(u) \otimes L_n(\lambda)]R(\lambda,u) \tag{14}$$

keeping in mind the ordering of ψ_{1n}, ψ_{2n} and interpreting (10) as a commutator we obtain

$$R(\lambda,u) = \begin{bmatrix} 1+c\,\frac{\lambda^2+u^2}{2(\lambda^2-u^2)} & 0 & 0 & 0 \\ 0 & 1 & \frac{c\lambda u}{\lambda^2-u^2} & 0 \\ 0 & \frac{c\lambda u}{\lambda^2-u^2} & 1 & 0 \\ 0 & 0 & 0 & 1+c\,\frac{\lambda^2+u^2}{2(\lambda^2-u^2)} \end{bmatrix} \tag{15}$$

It is interesting to note that the form of quantum R-matrix follows the same rule as that of the NLS equation, i.e., if one thinks of 'c' to be proportional to $\hbar$, (15) is true for calculation up to Δ .

2c Commutation Rule for the Scattering Data

Since written in discrete variable, the space part of the Lax equation becomes

$$\psi_{n+1} = L_n\psi_n \ . \tag{16}$$

One can interpret L_n as the transfer matrix over one lattice site.

We define the transition matrix

$$T(n,m/\lambda) = L_n(\lambda)L_{n-1}(\lambda) \ \ldots \ L_m(\lambda) \tag{17}$$

and the monodromy matrix

$$T(\lambda) = L_n(\lambda)L_{n-1}(\lambda)\ldots L_1(\lambda)$$

$$= \prod_{i=1}^{n} L_i(\lambda) \tag{18}$$

So from the basic relation (14) we get

$$R(\lambda,u)(T(\lambda)\otimes T(u)) = (T(u)\otimes T(\lambda))\, R(\lambda,u) \ . \tag{19}$$

We define

$$T(\lambda) = \begin{pmatrix} a(\lambda) & \bar{b}(\lambda) \\ b(\lambda) & -\bar{a}(\lambda) \end{pmatrix} \tag{20}$$

So coupling (19) we obtain all the commutation rules for $a,b,\bar{a},\bar{b}$. For example,

$$b(u)\bar{a}(\lambda) = \alpha(\lambda,u)a(u)b(u) - \beta(\lambda,u)\bar{a}(u)b(\lambda)$$

$$\bar{a}(\lambda)\bar{b}(u) = \alpha(\lambda,u)\bar{b}(u)\bar{a}(\lambda) - \beta(\lambda,u)\bar{b}(\lambda)\bar{a}(u) \tag{21}$$

where

$$\alpha(\lambda,u) = 1 + \frac{c(\lambda^2 + u^2)}{2(\lambda^2 - u^2)}$$

$$\beta(\lambda,u) = \frac{c\lambda u}{\lambda^2 - u^2}$$

2d Construction of the Eigenstates

The eigenstates of the quantized system can be constructed by starting with a postulated vacuum. Let us designate the vacuum state by

$$\psi_{2n} | 0 \rangle = 0 .$$

Then the operator L_n acts on $|0\rangle$ as a triangular matrix, i.e.,

$$L_n(\lambda) \; | \; 0\rangle = \begin{pmatrix} 1 - \frac{i\Delta\lambda^2}{2} & -\sqrt{c}\;\lambda\Delta\psi_{1n} \\ 0 & 1+ \frac{i\Delta\lambda^2}{2} \end{pmatrix} |0\rangle \tag{22}$$

From the property of triangular matrices we immediately get

$$T(\lambda) \; |0\rangle = \prod_{i=1}^{n} L_i(\lambda) \; | \; 0\rangle$$

$$= \begin{pmatrix} (1 - \frac{i\Delta\lambda^2}{2})^n & \sigma(\lambda) \; \psi_{1n} \\ 0 & (1 + \frac{i\Delta\lambda^2}{2})^n \end{pmatrix} |0\rangle \tag{23}$$

Making n very large so that $\Delta = L/N^-$ is small we get in the limit

$$T(\lambda) \; |0\rangle = \begin{pmatrix} e^{-iL\lambda^2/2} & \sigma(\lambda)\psi_{1n} \\ 0 & e^{iL\lambda^2/2} \end{pmatrix} |0 \rangle \tag{24}$$

So ψ_{1n} is to be used as the creation operator and also we get

$$a(\lambda) \quad |0 \rangle = e^{-iL\lambda^2/2} \quad |0 \rangle$$

$$\bar{b}(\lambda) \quad |0 \rangle = \sigma(\lambda) \; |0 \rangle$$

$$b(\lambda) \quad |0 \rangle = 0$$

$$\bar{a}(\lambda) \quad |0 \rangle = -e^{iL\lambda^2/2} \quad |0 \rangle \tag{25}$$

as the basic relations to be used in the construction of the algebraic Bethe Ansatz.

We now define a series of physical states of the form [9]

$$\Omega_1(\lambda_1) = \bar{b}(\lambda_1) \; |0\rangle$$

$$\Omega_2(\lambda_1\lambda_2) = \bar{b}(\lambda_1)\bar{b}(\lambda_2) \quad |0\rangle$$

$$\Omega_n(\lambda_1 \; ... \; \lambda_n) = \bar{b}(\lambda_1) \; ... \; \bar{b}(\lambda_n) \quad |0 \rangle . \tag{26}$$

To ascertain the eigenmomenta and eigenenergies of Ω_1, Ω_2, ..., we operate with a(u) and $\bar{a}$(u) on Ω_1, Ω_2,..., and utilize the commutation rules (21) to shift a(u) so as to operate on $|0\rangle$. Since in the sequel we define the Hamiltonian by Tr T(λ) = a(u) + $\bar{a}$(u), we demand $\Omega_1,\Omega_2,\ldots$, to be an eigenstate of a(u) + $\bar{a}$(u) which leads to the following equations determining the eigenmomenta and energy eigenvalues of the two particle state

$$[a(u) + \bar{a}(u)]\Omega_2(\lambda_1,\lambda_2) = [\alpha(\lambda_1,u)\alpha(\lambda_2,u)\, e^{-iu^2L/2}$$

$$- \alpha(u\lambda_1)\alpha(u\lambda_2)e^{iu^2L/2}]\Omega_2(\lambda_1\lambda_2) \qquad (27)$$

along with the condition

$$e^{i\lambda_1^2L} = \frac{\beta(\lambda_1,u)\alpha\,(\lambda_2\lambda_1)}{\beta(u,\lambda_1)\;\alpha(\lambda_1\lambda_2)}$$

$$e^{i\lambda_2^2L} = \frac{\alpha(\lambda_1,u)\beta(\lambda_2,u) - \beta(\lambda_1,u)\;\beta(\lambda_2\lambda_1)}{\alpha(u,\lambda_1)\beta(u,\lambda_2) - \beta(u,\lambda_1)\beta\;(\lambda_1\lambda_2)}$$

By using the expressions for $\alpha\,(\lambda,u)$ and $\beta\,(\lambda,u)$ we get

$$e^{i\lambda_1^2L} = \frac{\lambda_2^2(c + 2) + \lambda_1^2(c - 2)}{\lambda_1^2(c + 2) + \lambda_2^2(c - 2)}$$

$$e^{i\lambda_2^2L} = \frac{\lambda_1^2(c + 2) + \lambda_2^2(c - 2)}{\lambda_2^2(c + 2) + \lambda_1^2(c - 2)}$$

The same construction can easily be extended to the n particle states and equations are transcendental equations determining the eigenvalues (momenta) λ_i of excited states.

2e Refinement of the Yang-Baxter Relation

The above calculations are confined to the periodic case n→ -L to +L. But for n → ∞ the commutation relations get changed due to the modified Yang-Baxter relation.

The expectation value of $L_n(\lambda) \otimes L_n(u)$ between the vacuum states are [10]

$$W(\lambda,u) = \langle 0|\, L_n(\lambda) \otimes L_n(u) \quad |0\rangle$$

Thus

$$W(\lambda,u) = \begin{bmatrix} 1-\frac{i\Delta}{2}(\lambda^2+u^2) & 0 & 0 & 0 \\ 0 & 1-\frac{i\Delta}{2}(\lambda^2-u^2) & -ic\lambda u\Delta & 0 \\ 0 & 0 & 1+\frac{i\Delta}{2}(\lambda^2-u^2) & 0 \\ 0 & 0 & 0 & 1+\frac{i\Delta}{2}(\lambda^2+u^2) \end{bmatrix} \tag{28}$$

The normalized monodromy matrix $T(\lambda)$ is defined by [8]

$$T(\lambda) = \lim_{N\to\infty} v(\lambda)^{-N} L_n(\lambda) \ldots L_{-N+1}(\lambda)v(\lambda)^{-N}$$

where

$$v(\lambda) = \begin{pmatrix} w_1 & 0 \\ 0 & w_2 \end{pmatrix} ; \quad w_1 = 1 - \frac{i\Delta}{2}\lambda^2 \quad w_2 = 1 + \frac{i\Delta}{2}\lambda^2 \tag{29}$$

The Yang-Baxter relation for $T(\lambda)$ is modified to

$$R_1(\lambda\ u)(T(\lambda)\otimes T(u)) = (T(u)\otimes T(\lambda)R_2(\lambda\ u) \tag{30}$$

where

$$R_1(\lambda u) = U_1(u\lambda)^{-1}R(\lambda u)U_1(\lambda u)$$

$$R_2(\lambda u) = U_2(u\lambda)R(\lambda u)U_2(\lambda u)^{-1}$$

$$U_1(\lambda u) = \lim_{N\to\infty} W(\lambda u)^{-N}\ (v(\lambda)^N \otimes v(u)^N)$$

$$U_2(\lambda u) = \lim_{N\to\infty} (v(\lambda)^N \otimes v(u)^N)W(\lambda u)^{-N}$$

We computed $R_1(\lambda\ u)$ and $R_2(\lambda\ u)$ as

$$R_1(\lambda u) = \begin{bmatrix} 1+\frac{c(\lambda^2+u^2)}{2(\lambda^2-u^2)} & 0 & 0 & 0 \\ 0 & 1-f(\lambda u) & \frac{c\lambda u}{(\lambda^2-u^2)}(1+h(\lambda u)) & 0 \\ 0 & \frac{c\lambda u}{(\lambda^2-u^2)} & 1-g(\lambda u) & 0 \\ 0 & 0 & 0 & 1+\frac{c(\lambda^2+u^2)}{2(\lambda^2-u^2)} \end{bmatrix}$$

$$R_2(\lambda u) = \begin{bmatrix} 1+\dfrac{c(\lambda^2+u^2)}{2(\lambda^2-u^2)} & 0 & 0 & 0 \\ 0 & 1-g(\lambda u) & \dfrac{c\lambda u}{(\lambda^2-u^2)}(1+h(\lambda u)) & 0 \\ 0 & \dfrac{c\lambda u}{(\lambda^2-u^2)} & 1-f(\lambda u) & 0 \\ 0 & 0 & 0 & 1+\dfrac{c(\lambda^2+u^2)}{2(\lambda^2-u^2)} \end{bmatrix} \tag{31}$$

where

$$f(\lambda u) = \left(\frac{c\lambda u}{(\lambda^2-u^2)}\right)^2 [1 + 2\pi i\delta(\lambda^2-u^2)(\lambda^2-u^2+i0)]$$

$$g(\lambda u) = \left(\frac{c\lambda u}{(\lambda^2-u^2)}\right)^2 [1 - 2\pi i\delta(\lambda^2-u^2)(\lambda^2-u^2-i0)]$$

$$h(\lambda u) = -\frac{2c\lambda u}{(\lambda^2-u^2)}\left[1-\left(\frac{c\lambda u}{(\lambda^2-u^2)}\right)^2\right][1 - i\pi\delta(\lambda^2-u^2)]$$

One can now substitute $R_1(\lambda u)$, $R_2(\lambda u)$ in (30) to calculate the refined commutation rules of the scattering data and construct the eigenstates.

2f Construction of the Discrete Hamiltonian

Since the method of quantum inverse transform effectively works on the discretized $C_n(-L,L)$, it is necessary to formulate a discrete form of the Hamiltonian from the trace of the matrix $T(\lambda)$ (actually the logarithmic derivative of the $\mathrm{Tr}T(\lambda)$ with respect to λ), i.e.,

$$H = \frac{\partial}{\partial\lambda} L_n \,\mathrm{Tr}T(\lambda) . \tag{32}$$

When the determinant vanishes, the L operator is transformed into the one dimensional projection operator, i.e., $\det|L_n| = 0$ at

$$\lambda = \pm(c\psi_{1n}\psi_{2n} + \frac{2i}{\Delta})^{\frac{1}{2}} \tag{33}$$

So at such a value of the spectral parameter, L_n is degenerate and one can write L_n as the direct product of two vectors in the form

$$(L_n)_{ik} = \alpha_i(\Delta)\,\bar{\beta}_k(\Delta) \tag{34}$$

with

$$\alpha_i(\Delta) = \begin{pmatrix} 2(1 - \frac{ic\Delta}{2}\psi_{1n}\psi_{2n}) \\ -\sqrt{2i\Delta c}\,(1 - \frac{ic\Delta}{2}\psi_{1n}\psi_{2n})^{\frac{1}{2}}\psi_{2n} \end{pmatrix}$$

$$\bar{\beta}_k(\Delta) = \left(2(1 - \frac{ic\Delta}{2}\psi_{1n}\psi_{2n}) \;,\; -\sqrt{2i\Delta c}\,(1 - \frac{ic\Delta}{2}\psi_{1n}\psi_{2n})^{\frac{1}{2}}\psi_{1n}\right)$$

Constructing the monodromy matrix in a standard manner and using the expression for $(L_n)_{ik}$ in (34) we get

$$\begin{aligned} H &= \frac{\partial}{\partial\lambda}\ln \mathrm{Tr}T(\lambda) \\ &= (-i\Delta)^{\frac{1}{2}}\sum_{n=1}^{N} 2\alpha_1(n-1)^{\frac{1}{2}} + \sum_{n=1}^{N}\frac{\alpha_1(n)^{\frac{1}{2}}\alpha_1(n-1)^{\frac{1}{2}}\{\alpha_1(n-1)^{\frac{1}{2}} + \alpha_1(n)^{\frac{1}{2}}\}}{(\alpha(n-1).\alpha(n))} \end{aligned} \tag{35}$$

where $\alpha(n-1).\alpha(n)$ denotes the dot product of two dimensional vectors. Equation (35) actually represents a generalization of the nearest neighbour interaction usually adopted in lattice systems.

2g Exact form of $L_n(\lambda)$ operator [11]

Until now we were working with the form of L_n which is correct upto the order of Δ. But it is possible to construct a form of $L_n(\lambda)$ which is valid for all values of Δ. A procedure to do so was suggested by Sklyanin, Korepin and others. We assume the corrected version of L_n written as

$$\mathcal{L}_n(\lambda) = \begin{pmatrix} 1 - \frac{i\Delta}{2}(c\psi_{1n}\psi_{2n} + \lambda^2) & -\Delta\sqrt{c}\,\lambda\psi_{1n}\, f(\Delta,\psi_1\psi_2) \\ -\Delta\sqrt{c}\lambda\psi_{2n}\, f(\Delta,\psi_1\psi_2) & 1 + \frac{i\Delta}{2}(c\psi_{1n}\psi_{2n} + \lambda^2) \end{pmatrix} \tag{36}$$

where f is a function of the arguments Δ, ψ_1, ψ_2 such that as $\Delta \to 0$, it goes to one. We now assume that this corrected form of $L_n(\lambda)$ will lead to the same classical r-matrix as in (13) so that if we again compute $\{\bar{L}_n(x) \otimes \bar{L}_n(u)\}$ we will obtain some equations determining the function f. In the present case we have

$$\begin{aligned} i\Delta\{\psi_{1n}\psi_{2n}\,,\,\psi_{1n}\,f\} &= -\psi_{1n}\,f \\ i\Delta\{\psi_{1n}\psi_{2n}\,,\,\psi_{2n}\,f\} &= \psi_{2n}\,f \end{aligned} \tag{37}$$

An easy solution to these equations is seen to be

$$f = (1 + \frac{c\Delta^2}{4}\psi_{1n}\psi_{2n})^{\frac{1}{2}} \text{ which} \to 1 \text{ as } \Delta \to 0 \tag{38}$$

2h Derivation of the Properties of the Scattering Data

In the component form we rewrite the space part of the Lax equation as

$$v_{1x} = \frac{i\lambda^2}{2} v_1 - \frac{ic}{2} \psi_1 v_1 \psi_2 + \lambda\sqrt{c}\, \psi_1 v_2$$

$$v_{2x} = \frac{i\lambda^2}{2} v_2 - \frac{ic}{2} \psi_1 v_2 \psi_2 + \lambda\sqrt{c}\, v_1 \psi_2 \qquad (39)$$

where all the terms are written in specific normal ordered forms.

We use the asymptotic behaviour of the jost solutions to convert these into integral equations which are

$$\chi_1(x,\lambda)e^{i\lambda^2 x/2} = -\int_{-\infty}^{\infty} d\bar{x}\theta(\bar{x} - x) \frac{ic}{2} \psi_1(\bar{x})\chi_1(\bar{x},\lambda) e^{i\lambda^2\bar{x}/2}\psi_2(\bar{x})$$

$$+ \int_{-\infty}^{\infty} d\bar{y}\theta(\bar{y} - x)\lambda\sqrt{c}\; e^{i\lambda^2\bar{y}} \psi(\bar{y})\, \chi_2(\bar{y},\lambda)e^{-i\lambda^2\bar{y}/2}$$

along with

$$\chi_2(x,\lambda)e^{-i\lambda^2 x/2} = 1 - \int_{-\infty}^{\infty} d\bar{y}\theta(\bar{y}-x)\frac{ic}{2}\psi_1(\bar{y})\, \chi_2(\bar{y})e^{-i\lambda^2\bar{y}/2}\psi_2(\bar{y})$$

$$+ \int_{-\infty}^{\infty} d\bar{x}\theta(\bar{x} - x)e^{i\lambda^2\bar{x}}\lambda\sqrt{c}\; \chi_1(\bar{x},\lambda)\, e^{i\lambda^2\bar{x}/2}\psi_2(\bar{x}) \qquad (40)$$

We now adopt the procedure of iterating these integral equations. The important point to observe from this is that each term on the right-hand side generated after iteration has one operator ψ_1 excess and so the interpretation as a creation operator for χ_1.

Similarly it is seen that $\chi_2(x,\lambda)$ contains at every stage the same number of ψ_1 and ψ_2 so is equivalent to the number operator.

3. Conclusion

In the above analysis we have discussed the problem of a new integrable system. The new structure of $r(\lambda,u)$ and $R(\lambda,u)$ are the important and distinctive feature of the problem. It is also shown that one can construct an exact form of the classical L_n-operator for all Δ. Bethe eigenstates are easy to construct along with discrete Hamiltonians describing the nearest or next to nearest neighbour interactions, suitable for the description of the model on the lattice. Lastly, the operator interpretation of the scattering data and the exact form of the commutation rules for the scattering data are deduced by proceeding to the $N \to \infty$ limit.

Acknowledgments

One of the authors (S.S.) would like to thank Professor Abdus Salam, the International Atomic Energy Agency and UNESCO for the hospitality at the International Centre for Theoretical Physics, Trieste, during the Spring College on Plasma Physics, 1989. She is also grateful to DST (Government of India) for a senior research fellowship in a project.

References

[1] B. Buti, **Chaotic Alfven Waves**, Lecture Notes (1989).
[2] M. C. Gutzwiller, J. Math. Phys. **8** (1967) 1979; **11** (1970) 1791; **12** (1971) 343.
[3] L. D. Faddeev, Sov. Sci. Rev. Math. Phys. **C1** (1987) 107.
[4] V. O. Tarasov, Theor. Math. Phys. **61** (1984) 1211.
[5] E. K. Sklyanin, Dok. Akad. Nauk SSSR **244** (1978) 1337.
[6] V. S. Gerdjekov and M. I. Ivanov, JINR Preprint E2-82-595; Y. S. Vaklev, J. Math. Phys. **30** (1989) 1744.
[7] E. Gutkin, Phys. Rep. **167** (1988) 1.
[8] V. E. Korepin, Comm. Math. Phys. **86** (1982) 391.
[9] M. Wadati, A. Kuniba and T. Konishi, Phys. Soc. Japan **54** (1988) 1710.
[10] M. Wadati and K. Sogo, Prog. Theor. Phys. **69** (1983) 431.
[11] N. M. Bogolybov and V. E. Korepin, Theor. Math. Phys. **66** (1986) 455.

Nonlinear Chemical Dynamics

P.K. Chattaraj

Department of Chemistry, Indian Institute of Technology,
Kharagpur 721 302, India

This article reviews three different interesting nonlinear chemical dynamical problems tackled by us. It includes scattering from rough surfaces, topological equivalence of pseudorotation in cyclopentane with soliton in a ring and quantum fluid density functional theory of time-dependent phenomena, e.g., ion-atom collisions or atom-field interactions.

1. Introduction

It may not be inappropriate to say that all chemical processes are intrinsically nonlinear. Some of them appear to be linear due to unavailability of adequate instruments to resolve the nonlinear admixture with largely linear effects. These nonlinear aspects are not very transparent if one resorts to the conventional quantum mechanics of these processes. In order that the wavefunction obeys the superposition principle the linearity is built-in in Schrödinger's mechanics. There are two alternatives to overcome this difficulty. First one being the inclusion of a nonlinear potential in the classical mechanical framework and the second is to adopt Madelung's viewpoint of the quantum mechanics in terms of the charge density and the current density which ought to be nonlinear because the Schrödinger's linear mechanics in terms of the wavefunction precludes the density mechanics from being linear.

In this review article we discuss three different nonlinear processes occurring in chemical physics. Section 2 presents the scattering from fractal lattices while the pseudorotation in cyclopentanes is described in Section 3. Section 4 deals with the quantum fluid density functional theory of time-dependent phenomena.

2. Scattering from Fractal Lattices

Fractal behaviour [1] of a surface lattice has been found to be important in the study of heterogeneous reaction dynamics, the catalytic properties of solid surfaces and gas-surface interactions. There

Research Reports in Physics Symmetries and Singularity Structures
Editors: M. Lakshmanan · M. Daniel

are certain real systems which behave as fractals, e.g., crushed glass, fume silica, graphite and colloidal systems. We have studied [2,3] the scattering of a structureless particle from a model fractal lattice where the interaction potential contains the information about the dimensionality of the solid surface. The potential may represent the interaction between the approaching atom and the solid surface or with the adsorbate atoms on a surface. The relevant quantity for this study is the T-matrix element.

$$T^{d_f}(\mathbf{k}',\mathbf{k}) = \langle \varphi_{\mathbf{k}'} | V^{d_f} | \psi_{\mathbf{k}}^{+d_f} \rangle \quad , \tag{1}$$

where V^{d_f} is the interaction potential labelled by the fractal dimension d_f of the surface and the scattered wavefunction $\psi_{\mathbf{k}}^{+d_f}$ is related to the incident wavefunction $\varphi_{\mathbf{k}}$ by the Lippmann-Schwinger equation,

$$\psi_{\mathbf{k}}^{+d_f}(\mathbf{r}) = \varphi_{\mathbf{k}}(\mathbf{r}) + \int G^{+}(\mathbf{r},\mathbf{r}')V^{d_f}(\mathbf{r}')\psi_{\mathbf{k}}(\mathbf{r}')d\mathbf{r}' \, , \tag{2}$$

where $G^{+}(\mathbf{r},\mathbf{r}')$ is the Green function for forward scattering. Equation (2) has been solved using average wavefunction method [4]. Our calculation starts with the construction of model fractal-like lattices comprising cantor-like sets of two different dimensions, viz., $d_f = \log 2 / \log 3$ and $d_f = \log 3 / \log 5$ in different generations. Knowing the surface geometry the interaction potential can be written within the pairwise interaction approximation as,

$$V^{d_f}(\mathbf{r}) = \sum_{i=1}^{n} v_i(\mathbf{r} - \mathbf{r}_i) \, . \tag{3}$$

For simplicity we consider spherically symmetric gaussian potential having the variance a at each atomic site where the atomic position $\mathbf{r}_i$ is governed by d_f. Scattering observables like differential cross-sections can easily be evaluated once the T-matrix is known. We have used a gaussian elimination technique for matrix inversion and 4-point Gauss-Legendre quadrature for numerical integration.

Figures 1a,b show the results for 16 (third generation) and 32 (fourth generation) atoms respectively for the lattice with the same fractal dimension ($d_f = \log 2/\log 3$). For different ak values different scales have been used for plotting. The increasing number of peaks as one goes from lower to higher energies for incoming waves, is expected because of diffraction effects.

Figure 2 depicts the logarithmic plot of the integrated cross-section (integrated over the scattering angle only) against the scaled

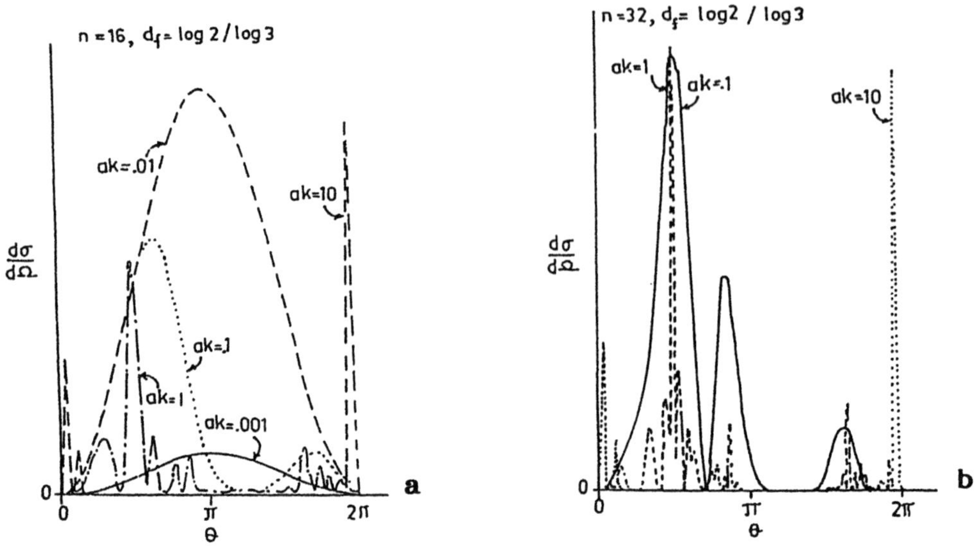

Fig. 1 Differential scattering cross-section ($d\sigma/d\Omega$), in units of $4\pi^2a^2$, plotted against the scattering angle (Θ), where 'a' is the variance of the gaussian potential $v = \exp(-r^2/2a^2)$ at each atomic site, and the fractal dimension of the lattice is log 2/log 3. The number of atoms is (a) 16 (ak = 0.001 to 10); and (b) 32 (ak = 0.1 to 10). The parameter ak is the scaled wave-vector. For different values of ak the scales are different.

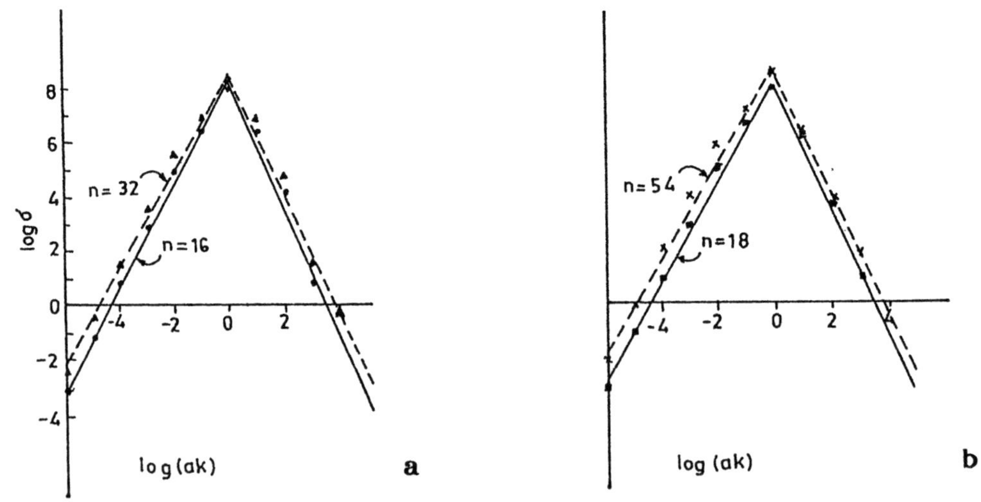

Fig. 2 Logarithm of integrated cross-section plotted against the logarithm of the wave-vector scaled with respect to the variance of the gaussian potential at each atomic site (a) d_f = log 2/log 3: slope at ak < 1 = 1.75 (n = 16), 1.80 (n = 32): slope for ak > 1 = -1.8 (n = 16), -1.72 (n = 32); (b) d_f = log 3/log 5: slope at ak < 1 = 1.75 (n = 18, 54): slope at ak > 1 = -2.08 (n = 18), -2.1 (n = 54), n is the number of atoms in the lattice.

wavenumber of the incoming wave for two different generations of two different fractal sets. The crossover in the linear slope when ak is unity is very significant. The linear slope confirms the fractal property of the cross-section data which means that the memory of fractal geometry of the surface is retained in the scattering observables through the interaction potential. The self-similarity is reflected in the near constancy of the slopes for the different generation lattices having same fractal dimension. Although the crossover from Euclidean to non-Euclidean dimensions is known for spin-lattice relaxation of proteins [5] corresponding physics is not very transparent as also in the present study.

3. Pseudorotation in Cyclopentanes and Solitons in a Ring

The phenomenon of pseudorotation in cyclopentane and its derivatives has been known for a long time [6]. It might be a natural guess that the most symmetric planar cyclopentane molecule with D_{5h} symmetry would be the most stable one. However, it has been observed that cyclopentane and most of its derivatives prefer to be in one of the two extremes, viz., envelope and half-chair or in any other conformation in between these two. In these two conformations the planar symmetry D_{5h} is reduced to C_s or C_2 respectively for the instantaneous planar C-C framework. In the envelope conformation there is a symmetry plane perpendicular to the average plane of the cyclopentane ring and hence its molecular point group is C_s where the molecule resembles an open envelope whose flap is pointing upward. On the other hand, the half-chair form belongs to C_2 point group because of the presence of the C_2 axis in the molecular plane. The name half-chair has been given owing to its similarity with the rigid chair form of cyclohexane having local tetrahedral symmetry around each carbon atom. The C-C-C-C dihedral angles are slightly different in chair and half-chair forms. It has been rather established that the singleton out of plane position in envelope conformation would be assumed by all the five carbon atoms one after another through the rotation of the puckering around the ring popularly known as pseudorotation. In order to study the pseudorotation dynamics we have mapped [7] it onto a soliton in a ring problem where a sine-Gordon field with the topology of a ring has been considered. The puckering rotating around the cyclopentane ring resembles the problem of a soliton (a twist or a kink) which has been confined to a ring due to topological restraints or due to the nonlinearities in the dynamics of the ring. Accordingly, a soliton in a ring may be considered as a theoretical model for pseudorotation due to the topological equivalence. The equation obeyed by the field is

$$\varphi_{xx} - \varphi_{tt} - m^2 \sin\varphi = 0 \tag{4}$$

with the periodic boundary condition

$$\varphi(x,t) = \varphi(x + 2\pi, t) . \tag{5}$$

Defining a new variable as

$$\tau = \left[\frac{m^2}{\omega^2 - 1}\right]^{\frac{1}{2}} S \tag{6}$$

$$= \gamma \mathbf{S} = \gamma(x - \omega t)$$

Eq. (4) can be transformed into

$$\varphi_{\tau\tau} = -\frac{\partial}{\partial\varphi}[V(\varphi)] \tag{7}$$

where

$$V(\varphi) = 1 - \cos\varphi . \tag{8}$$

In Eq. (6), ω is the angular velocity with which the soliton moves around the ring. It is interesting to note that Eq. (7) can be considered as the Newton's equation of motion for a pseudo-particle moving in the periodic potential given by Eq. (8). The motion is unbounded and no solution satisfying the boundary condition (5) is possible if the energy E_p of the pseudo-particle is greater than 2. However, if E_p is less than 2 then the motion can be periodic, i.e.,

$$\varphi(\tau + T) = \varphi(\tau) \tag{9}$$

where T is the period.
The solutions of Eq. (7) with the periodic boundary condition (9) have been found to be

$$F(\varepsilon, k) = \tau \tag{10}$$

and

$$T = 4K(k) , \tag{11}$$

where $F(\varepsilon, k)$ and $K(k)$ are incomplete and complete elliptic integrals of the first kind respectively [8],

$$\varepsilon = \sin^{-1}[\sin(\varphi/2)/k] \text{ and } E_p = 2k^2.$$

The model described here can be quantized as follows, writing the energy of the field configuration as

$$\mathbb{E}[\varphi] = \int_0^{2\pi} ds\ \{(\varphi_x^2 + \varphi_t^2)/2 + m^2 V(\varphi)\} \qquad (12)$$

and invoking the conservation of the energy of the pseudoparticle. This gives

$$\mathbb{E}[\varphi] = 4\pi m^2 - 4\pi\gamma^2(\omega^2+1)\mathrm{dn}^4(\pi\gamma/4) + 4\gamma\{E(2\pi\gamma) - E(0)\}\ , \qquad (13)$$

where dn(k) and E(k) are Jacobian elliptic function and incomplete elliptic integral of the second kind respectively [8]. Define the momentum conjugate to ω as

$$J(\omega) = \frac{\partial\mathbb{E}(\omega)}{\partial\omega} \qquad (14)$$

and expand $J(\omega)$ as the Maclaurin's series in ω . Applying the calculus of residue technique the above series can be inverted as follows:

$$\omega = \sum_n J_n J^n \qquad (15)$$

where

$$J_n = \frac{1}{n!}\left[\frac{d^{n-1}}{d\omega^{n-1}}\left(\frac{\omega}{J(\omega)}\right)^n\right]_{\omega=0} . \qquad (16)$$

Now, imposing the quantization condition for J we have

$$\mathbb{E}_M[\varphi] = \mathbb{E}(\sum_n J_n(M\hbar)^n), \quad M = 0, \pm1, \pm2, \ldots. \qquad (17)$$

Thus, the topological mapping between the pseudorotation of the puckering around the cyclopentane or any cycloalkane ring and the rotational motion of the soliton provides additional insights into the former, particularly regarding the quantized pseudorotation energy levels and energy necessary for having a transition between any two such levels which may be monitored spectroscopically.

4. Electron Density Dynamics of Time-Dependent Processes

Density functional theory (DFT) has become very successful [9-11] in ground state electronic structure, binding, property and reactivity calculations in many-fermion systems including atoms, molecules and solids. The popularity of DFT stems from its physical, mathematical, computational and conceptual simplicity because it states that a real-valued 3-D quantity, viz. the electronic charge density contains all information about a many-fermion system. Its application to time-dependent (TD) problems and problems involving excited states is, however, relatively new.

The joint use of TDDFT and quantum fluid dynamics (QFD) has been found to be useful in tackling dynamical problems where electronic charge and current densities together describe any system or a process. A quantum fluid density functional theory (QFDFT) has been developed [12-14] through an amalgamation of DFT and QFD as well as from the stochastic interpretation of quantum mechanics [15]. The QFDFT has been applied in studying high energy ion-atom collisions [12] and atom-field interactions [14].

The two basic equations in QFD are continuity equation and the equation of motion. In terms of the charge density $\rho(\mathbf{r},t)$ and a velocity potential $\chi(\mathbf{r},t)$ (the current density $\mathbf{j}(\mathbf{r},t) = \rho(\mathbf{r},t)\nabla\chi(\mathbf{r},t)$) they can be written as follows:

$$\frac{\partial\rho}{\partial t} + \nabla\cdot(\rho\nabla\chi) = 0 . \tag{18}$$

$$\frac{\partial\chi}{\partial t} + \frac{1}{2}(\nabla\chi)^2 + \frac{\delta G[\rho]}{\delta\rho} + \int \frac{\rho(\mathbf{r}',t)}{|\mathbf{r} - \mathbf{r}'|}\, d\mathbf{r}' + v(\mathbf{r},t) = 0 . \tag{19}$$

Here the universal functional $G[\rho]$ comprises kinetic and exchange-correlation energy functionals.

The external potential $v(\mathbf{r},t)$ has the following forms for two different time-dependent processes considered here, viz.

(a) For proton-atom collisions, $v(\mathbf{r}, t)$ is given by

$$v(\mathbf{r},t) = -\left(\frac{Z}{r} + \frac{1}{|\mathbf{R}(t) - \mathbf{r}|}\right) \tag{20a}$$

where Z is the nuclear charge of the atom and R(t) is the Coulomb trajectory on which the two nucleii move.

(b) For an atom in a laser field, within the dipole approximation, $v(\mathbf{r},t)$ can be written as

$$v(\mathbf{r},t) = -\left(\frac{Z}{r} - \varepsilon_t \sin(\omega t)\, z\right) \tag{20b}$$

where ω is the frequency of the electromagnetic wave and the field amplitude ε_t is written as

$$\begin{aligned} \varepsilon_t &= \varepsilon_0 t/t' \text{ for } 0 \le t \le t' \\ &= \varepsilon_0 \text{ otherwise} \end{aligned} \tag{20c}$$

where ε_0 is the maximum field amplitude, time t is a positive semi-definite parameter and t' is the switch-on time for the laser.

Now, write a complex-valued 3D function $\varphi(\mathbf{r},t)$, in its polar form, in terms of two real-valued functions $\rho(\mathbf{r},t)$ and $\chi(\mathbf{r},t)$ as

$$\varphi(\mathbf{r}, t) = \rho^{1/2}(\mathbf{r},t)\exp[i\chi(\mathbf{r},t)] . \tag{21}$$

Eliminating χ from Eqs. (18) and (19) and making use of Eq. (21) we obtain the following generalized nonlinear Schrödinger equation (GNLSE) in 3D space:

$$[-\frac{1}{2}\nabla^2 + v_{eff}(\mathbf{r},t)]\,\varphi(\mathbf{r},t) = i\,\frac{\partial\varphi(\mathbf{r},t)}{\partial t} \tag{22}$$

where the effective one-body, nonlocal, nonlinear pulsating potential $v_{eff}(\mathbf{r},t)$ is given as

$$v_{eff}(\mathbf{r},t) = \frac{\delta T_{NW}[\rho]}{\delta\rho} + \frac{\delta E_{xc}[\rho]}{\delta\rho} + \int \frac{\rho(\mathbf{r}',t)}{|\mathbf{r} - \mathbf{r}'|}\, d\mathbf{r}' + v(\mathbf{r},t). \tag{23}$$

In obtaining Eq. (23), the kinetic energy (KE) functional has been considered as

$$T[\rho] = T_W[\rho] + T_{NW}[\rho], \tag{24a}$$

where $T_W[\rho]$ is the Weizsäcker energy functional given by

$$T_W[\rho] = \frac{1}{8}\int \frac{\nabla\rho\cdot\nabla\rho}{\rho}\, d\mathbf{r} \tag{24b}$$

and $T_{NW}[\rho]$ is the corresponding non-Weizsäcker part. Equation (22) has been solved in order to study the high-energy (25 keV) proton-neon atom head-on collision [12] as well as the interaction of a helium atom with a laser field [14]. Considering the azimuthal symmetry for both of the above problems the GNLSE is written in cylindrical polar coordinates ($0 \le \tilde{\rho} \le\infty$, $-\infty \le z \le \infty$, $0 \le\tilde{\varphi} \le 2\pi$) and the integration over the azimuthal angle $\tilde{\varphi}$ has been carried out analytically. For convenience, we also use the following transformations

$$Y = \tilde{\rho} \tag{25a}$$

$$\tilde{\rho} = x^2 \tag{25b}$$

such that Eq. (22) becomes

$$\left[\frac{3}{4x^3}\frac{\partial y}{\partial x} - \frac{1}{4x^2}\frac{\partial^2 y}{\partial x^2} - \frac{\partial^2 y}{\partial z^2}\right] - \left[\frac{1}{x^4} - 2\,v_{eff}\right] y = 2i\,\frac{\partial y}{\partial t} . \tag{26}$$

This is an initial boundary value problem which has been solved by

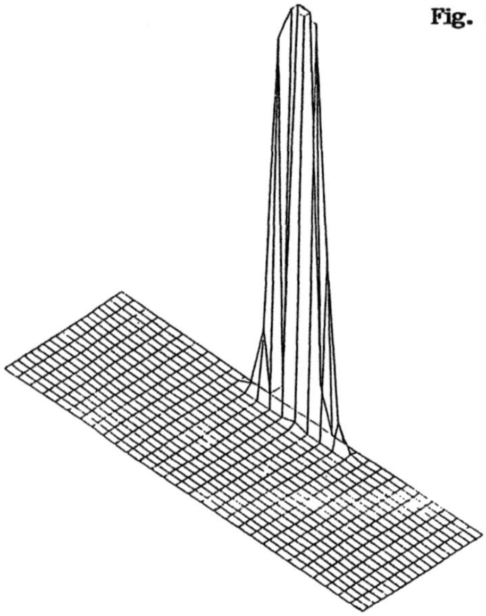

Fig. 3

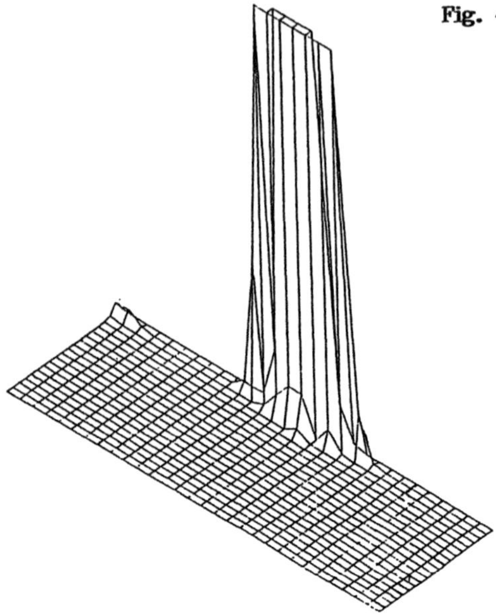

Fig. 4

Fig. 3 Perspective plot (a.u.) of $0 \leq \rho(\mathbf{r},t) \leq 50$ in the proton-neon atom scattering system (head-on collision, proton energy 25 keV, t = 0.08 a.u.) in cylindrical polar coordinates ($\tilde{\rho}$, z). The target neon nucleus is at (0,0) and the proton is approaching from the left along the $\tilde{\rho} = 0$ direction of z.

Fig. 4 Perspective plot (a.u.) of $0 \leq |\mathbf{j}(\mathbf{r},t)| \leq 50$ in the proton-neon atom scattering system in cylindrical polar coordinates. See caption of Fig. 3 for other details.

using an alternating direction implicit scheme for the proton-atom collision problem and by the leapfrog scheme for the atom-field interaction problem. Both the schemes are stable, convergent and consistent with the original partial differential equation [16]. The boundary conditions used for this purpose are:

At $t = 0$, $y(x,z)$ is known, $\forall\ x, z$

(input density, near-Hartree-Fock [17], Dirichlet data) (27)

$$Y(0,z) = 0 = Y(\infty, z), \quad \forall\ z,t$$
$$Y(x, \pm\infty) = 0, \quad \forall\ x,t\ .$$

Figures 3-5 depict the perspective plots of TD ρ, $\mathbf{j}$ and v_{eff} respectively in the ($\tilde{\rho}$, z) mesh, at t = 0.08 a.u. for proton-neon collision problem. The entire potential surface oscillates as time progresses. Collective nonlinear charge oscillations are prominent in the plots of ρ and $\mathbf{j}$. The nonvanishing $\mathbf{j}$ indicates the mixing of excited states with the ground state of neon atom as the proton approa-

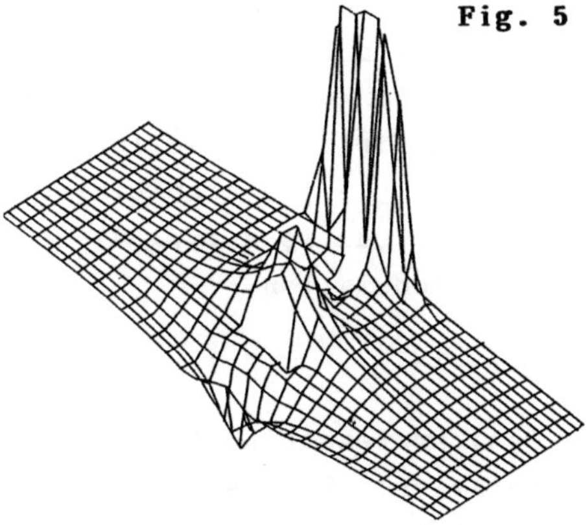

Fig. 5 Perspective plot (a.u.) of $-50 \leq v_{eff}(\mathbf{r},t) \leq 50$ in the proton-neon atom scattering system in cylindrical polar coordinates. See caption of Fig. 3 for other details.

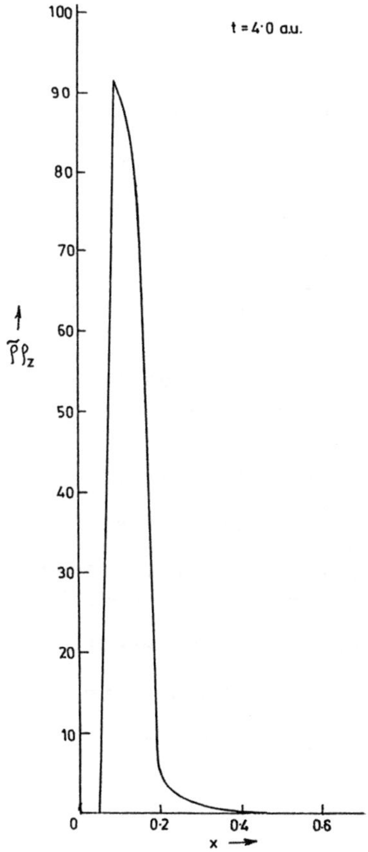

Fig. 6 Plot (a.u.) of $\tilde{\rho}\rho_z$ against $\sqrt{\tilde{\rho}}$ for helium atom interacting with a laser field of unit intensity at t = 4 a.u.

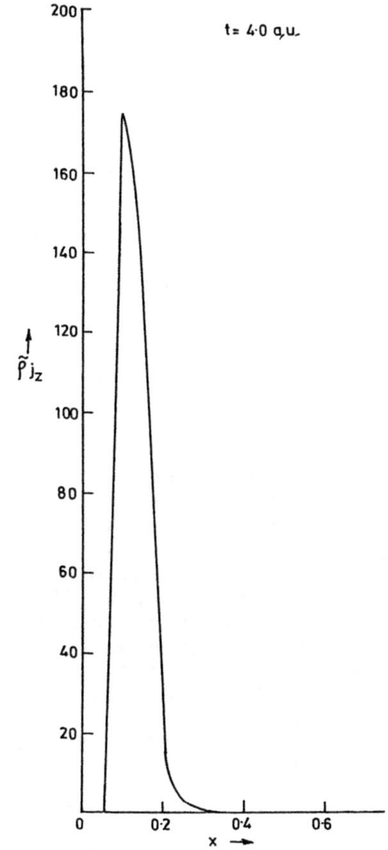

Fig. 7 Plot (a.u.) of $\tilde{\rho}j$ against $\sqrt{\tilde{\rho}}$ for helium atom interacting with a laser field of unit intensity at t=4 a.u.

ches. Time evolution of any dynamical observable of the colliding systems can be studied knowing $\rho(\mathbf{r},t)$ and $\mathbf{j}(\mathbf{r},t)$ at any instant of time.

Quantities $\tilde{\rho}\rho_z$ and $\tilde{\rho}j_z$ (any z-averaged function $u_z = 2\pi\int_{-\infty}^{\infty}\tilde{u}(x,z,t)dz$) are plotted against x, at t = 4 a.u., for helium-field interaction problem and presented in Figs. 6 and 7 respectively. The accumulation of both charge and current densities near the nucleus is a typical nonlinear phenomenon originating from the self-modulation.

Acknowledgments

The author is thankful to Professor B. M. Deb, Professor K. L. Sebastian and Dr Harjinder Singh for collaboration. Thanks are also due to Professor M. Lakshmanan for providing with the opportunity of participating in the present workshop.

References

[1] B. B. Mandelbrot, **The Fractal Geometry of Nature** (Freeman, San Fransisco, 1983).
[2] H. Singh and P. K. Chattaraj, Proc. Indian Acad. Sci. (Chem. Sci.) **99** (1987) 47, 207.
[3] H. Singh and P. K. Chattaraj, Phys. Lett. **A128** (1988) 355.
[4] H. Singh, D. K. Dacol and H. Rabitz, J. Chem. Phys. **84** (1986) 1373, 1852.
[5] K. N. Srivastava, J. Phys. **C20** (1987) L381.
[6] E. Kilpatrick, K. S. Pitzer and R. Spitzer, J. Am. Chem. Soc. **69** (1947) 2483.
[7] P. K. Chattaraj and K. L. Sebastian, to appear.
[8] P. F. Byrd and M. D. Friedman, **Handbook of Elliptic Functions for Scientists and Engineers** (Springer-Verlag, Berlin, 1971).
[9] S. K. Ghosh and B. M. Deb, Phys. Rep. **92** (1982) 1.
[10] B. M. Deb and P. K. Chattaraj, to appear.
[11] N. H. March and B. M. Deb,eds., **Single-Particle Density in Physics and Chemistry** (Academic, London, 1987).
[12] B. M. Deb and P. K. Chattaraj, Phys. Rev. **A39** (1989) 1696.
[13] B. M. Deb and P. K. Chattaraj, in **Solitons: Introduction and Applications**, ed. M. Lakshmanan (Springer-Verlag, Berlin, 1988); B. M. Deb and P. K. Chattaraj, Chem. Phys. Lett. **148** (1988) 550.
[14] P. K. Chattaraj, to appear.
[15] E. Nelson, **Quantum Fluctuations** (Princeton University Press, New Jersey, 1985).
[16] P. K. Chattaraj, S. Rao Koneru and B. M. Deb, J. Comp. Phys. **72** (1987) 504.
[17] E. Clementi and C. R. Roetti, At. Data Nucl. Data Tables **14** (1974) 174.

Dynamics of Solitons on ^{4}He Films

V.M. Nandakumaran and J. Sreekumar

Department of Physics, Cochin University of Science and Technology, Cochin 682022, India

In this paper we present a numerical study of the time evolution of local density waves in thin ^{4}He films. We also consider surface wave propagation in a saturated film and show that in the small amplitude regime the dynamics is described by the KP equation. The phenomenon of two-soliton resonances is also discussed.

The study of the propagation and the interaction of solitons in superfluid ^{4}He films has attracted considerable attention in recent years. Analytical and numerical methods have been applied to study localised disturbances both on thin films (thickness $\sim 10^{-7}$ cm) and on the so-called saturated films (thickness $\sim 10^{-6}$ cm). Huberman [1] was the first to suggest that nonlinear local density fluctuations in very thin ^{4}He films can travel unattenuated for large times. Nakajima et al. [2] considered one dimensional solitary waves in the ^{4}He films and showed that they satisfied the Korteweg-deVries (KdV) equation. Biswas and Warke [3] starting from the phenomenological equation of Rutledge et al. [4] derived the KdV equation for superfluid density fluctuations and confirmed the predictions about the existence of solitons on ^{4}He films. Later they generalised these results [5] to quasi-two-dimensional wave propagation (where the direction of propagation was chosen to be along the x-axis and the y-dependence was assumed to be weak) and obtained the Kadomtsev-Petviashvili (KP) equation. Using these results we have [6] studied the phenomenon of two-soliton resonance of the KP equation for the superfluid surface density fluctuations and derived the velocity of the resonant solitons.

Kurihara [7] was the first to study the dynamics of wave propagation on ^{4}He films by retaining the full nonlinearity of the van der Waals potential between the film and the substrate. By numerical studies he showed that there exists soliton-like localised excitations-quasi-solitons--which have a high degree of stability under collisions. These quasi-solitons had profiles which are localised along only one of the directions.

This lecture essentially consists of two parts. In the first part we consider the large amplitude local density fluctuations on thin ^{4}He films and present numerical studies of the time evolution of these density fluctuation profiles. In the second part we concentrate on saturated films and look at the fluctuations in thickness of the films. In particular we study the problem of soliton resonances [8]. This is in contrast to the earlier work [6] where only the density fluctuations were considered and the surface deformations were completely neglected.

Density Waves on Thin He Films

In this section we investigate the time evolution of localised density fluctuations on a thin film of ^{4}He by taking into consideration the nonlinearity of the van der Waals potential. The relevant equations are written in a dimensionless form. The analysis is done numerically as an initial value problem where an arbitrary initial profile is used. This is essentially an extension of the work of Kurihara [7] in which we assume that the initial profile is localised in both x and y-directions.

We begin with the equation of motion for the condensate wave function ψ [4]

$$i\hbar \frac{\partial\psi}{\partial t} = -\frac{\hbar^2}{2m} \nabla^2\psi - \frac{A\psi}{(a+|\psi|^2)^3} - \mu\psi - B\psi\nabla^2|\psi|^2 , \tag{1}$$

where m is the mass of the ^{4}He atoms, A and a are the constants of the van der Waals interaction, μ is the chemical potential and B is the surface tension. Recently, Radha Balakrishnan et al. [9] have derived a nonlinear evolution equation for the condensate wave function from a microscopic model of bosons with hard-core repulsion together with an attractive nearest neighbour interaction. Equation (1) is obtained as an approximation to their equation, on neglecting certain nonlinear terms. We put B = 0 since we are considering a monolayer film. We look for solution of the form

$$\psi(x,y,t) = \rho_s^{1/2}(x,y,t)\, e^{i\varphi(x,y,t)} \tag{2}$$

where ρ_s is the superfluid density. From equation (1) we get the equation of continuity

$$\frac{\partial\rho_s}{\partial t} + \nabla\cdot\mathbf{J}_s = 0 \tag{3}$$

where $J_s(x,y,t) = \mathrm{Re}[(\hbar/im)\psi^*\nabla\psi\,]$ is the current density.

For doing numerical studies we transform equation (1) to a dimensionless form. For this we express ψ in terms of its equilibrium value ψ_0 obtained from the relation

$$\mu + \frac{A}{(a+|\psi_0|^2)^3} = 0 . \qquad (4)$$

We fix the equilibrium value of the superfluid thickness as

$$d_0 = \psi_0^2/a . \qquad (5)$$

Define

$$k = (2mW/\hbar^2)^{1/2} \qquad (6)$$

$$\omega = W/\hbar \qquad (7)$$

where

$$W = A/a^3(1 + d_0)^3 \qquad (8)$$

is the van der Waals energy.

Setting

$$\chi = \psi/\psi_0 , \qquad (9)$$

$$\xi = kx , \qquad (10)$$

$$\eta = ky , \qquad (11)$$

$$\tau = \omega t , \qquad (12)$$

we can write equation (1) as

$$i \frac{\partial\chi}{\partial\tau} = - \frac{\partial^2\chi}{\partial\xi^2} - \frac{\partial^2\chi}{\partial\eta^2} - \left(\left[\frac{1 + d_0}{1 + d_0|\chi|^2}\right]^3 - 1\right) \chi . \qquad (13)$$

It is worth mentioning that an exact one-soliton solution of equation (13) in (1+1)-dimensions has been found analytically by Kurihara [10]. To solve equation (13) numerically we assume that initially the superfluid density is altered locally. This could be done, for example, by heating the film, locally. For simplicity we look for solutions propagating along the ξ-axis. The size of the film (=100 along the ξ direction) is chosen arbitrarily such that its size is larger than the characteristic size of the localised excitations. Equation (13) is treated as an initial value problem. At t = 0 we assume that the whole superfluid is at rest; that is, we choose the initial value of the phase of the wave to be constant throughout the film. It is found that the dynamics of the film is independent of the actual value of this constant.

Large Amplitude Fluctuations

We now consider the time evolution of the superfluid density

$$\rho(\xi, \eta, \tau) = |\psi(\xi, \eta, \tau)|^2 - 1. \quad (14)$$

For numerical analysis we consider two distinct initial profiles for $\rho(\xi, \eta, 0)$. In the first case we choose a squared secant hyperbolic profile. This choice is of special importance. We are essentially looking at the fate of the KP solitons under the full nonlinearity. In the second case we choose a rounded rectangular form. The time evolution in these two cases has different characteristics. In both we have imposed the periodic boundary conditions

$$\rho(100, \eta, \tau) = \rho(0, \eta, \tau) . \quad (15)$$

The numerical results for the first case are shown in Figs.1 and 2. Figure 1 shows the time evolution of the superfluid density for the initial profile

$$\rho(\xi, \eta, 0) = \rho_0 \mathrm{sech}^2[\{(\xi - 25)/3\} + \{(\eta - 50)/55\}] . \quad (16)$$

The parameters chosen are $d_0 = 1$ and $\rho_0 = 0.2$. The velocity of the fluctuations are plotted in Fig. 2. The numerical experiments carried out for other values of the parameters showed similar results. It is seen that the single peak of the initial profile splits up into two peaks ("solitons") and travel in opposite directions. The two peaks more or less presume their identity after interactions.

In the second case we use the rounded rectangular form

$$\rho(\xi,\eta,0) = \rho_0 \frac{1}{2}\left\{1 - \tanh\left[\{(\xi - 25)/1.5\}^2 + \{(\eta - 50)/55\}^2 - 4\right]\right\} \quad (17)$$

as the initial profile. Again we choose $d_0 = 1$ and $\rho_0 = 0.2$. The resulting temporal developments are plotted in Figs. 3 and 4.

As in the previous case, the initial profile splits into two. However, as the resulting waves travel in opposite directions they continously emit secondary waves. This behaviour continues even after a few collisions. This could be explained in the following manner. The initial disturbance, by continuously emitting secondary waves, adjusts its shape so as to reach a stable form which can then continue unattenuated for a long time. The emission is more profuse the larger the deviations of the initial profile from the ideal soliton profile. This can also be seen in the first case where, however, the emission of the secondary waves is extremely weak. The study using a narrow

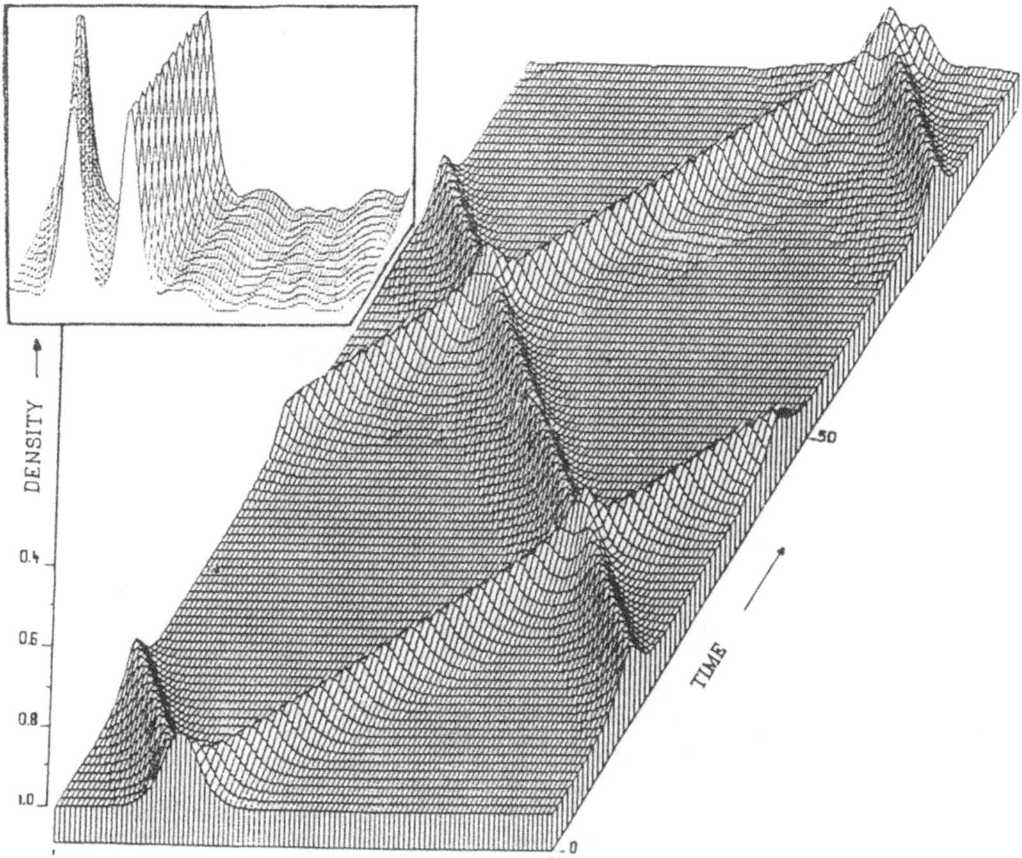

Fig. 1

Time evolution of the amplitude of the superfluid density fluctuations occurring along the ζ-direction, in arbitrary units, for the squared secant hyperbolic initial profile. A magnified version of the profile at large times is drawn in the inset to display the emitted waveform.

Gaussian profile confirms this where the emission of the secondary wave is not as profuse as in the rounded rectangular form.

We have shown numerically that even under strong nonlinearity the ^{4}He film admits composite solitons.

In the next section we consider fluctuations in thickness in a saturated ^{4}He film.

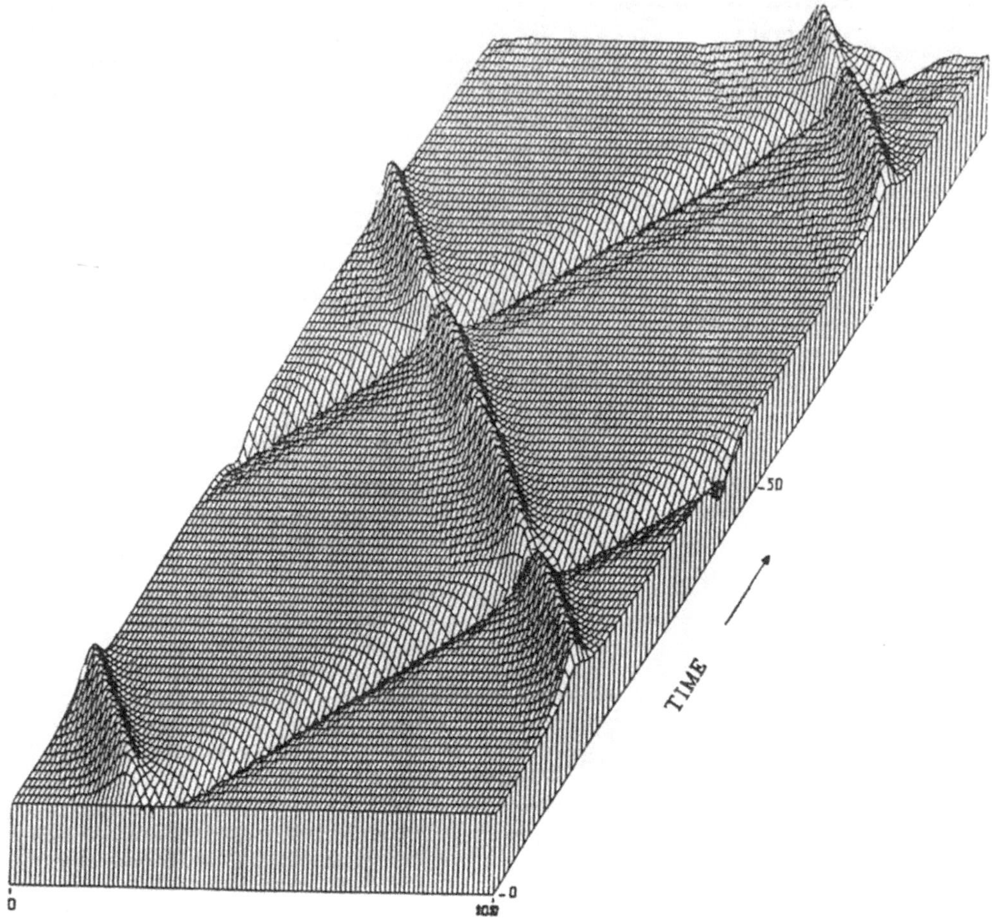

Fig. 2
The superfluid velocity corresponding to the density fluctuation plotted in Fig. 1, in arbitrary units

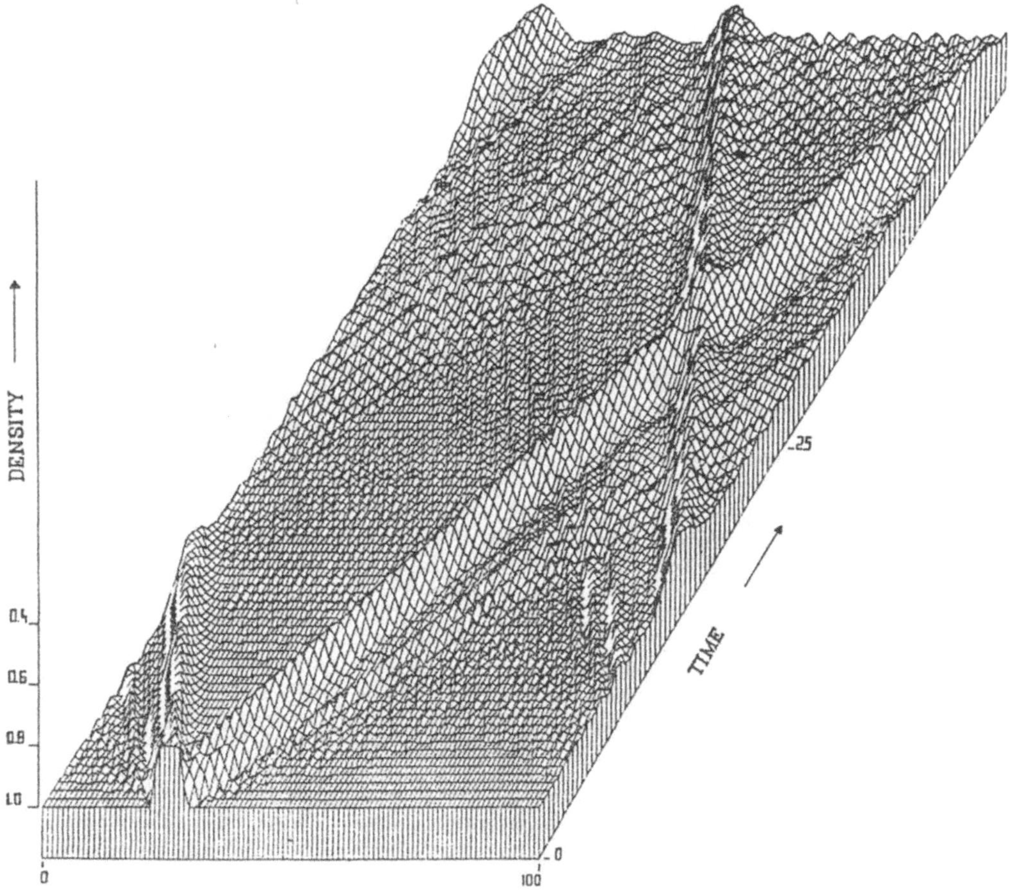

Fig. 3

Superfluid density fluctuations as in Fig. 1, but with a rounded rectangular form as the initial profile

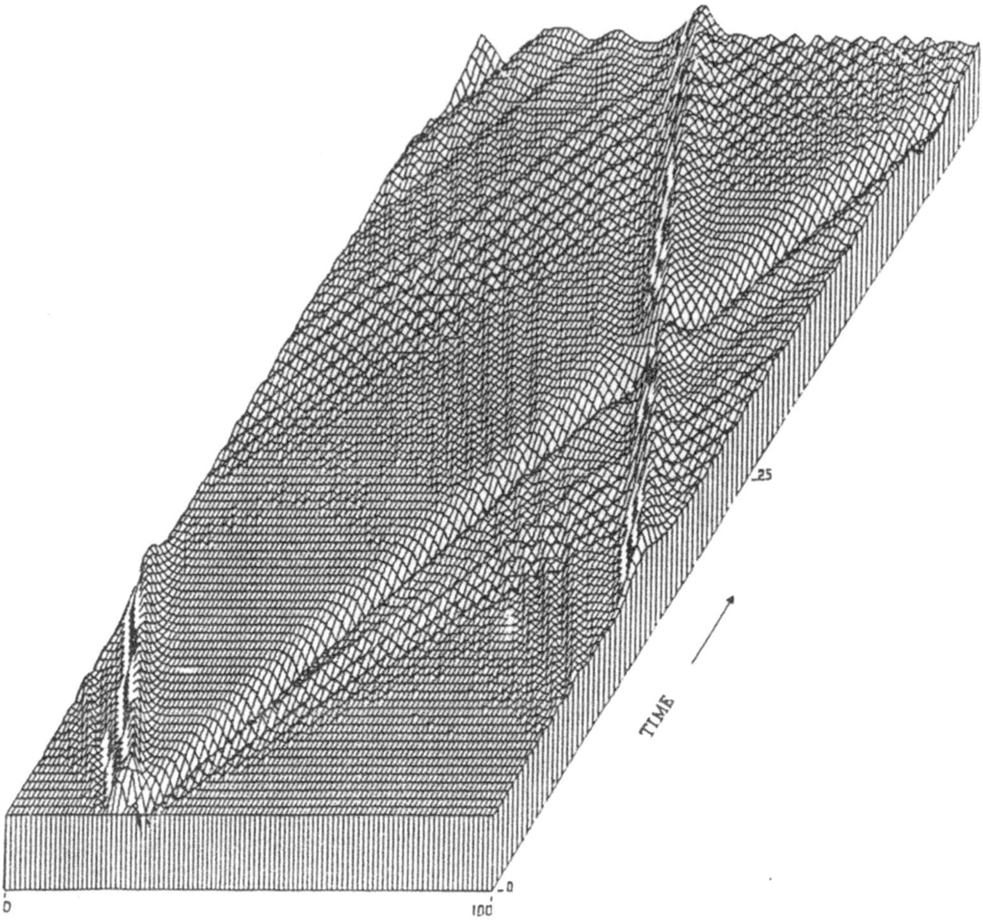

Fig. 4
Superfluid velocity corresponding to the density fluctuations plotted in Fig. 3, in arbitrary units

Surface Waves on Saturated Films

In this section we consider the propagation of surface waves on saturated ^4He films [11]. Here one has to include the effects of surface tension which is generally ignored for every thin films. We consider the x and y axes to be lying parallel to the substrate. Geometry of the system is depicted in Fig. 5.

In what follows we obtain the equation of motion for the surface displacement. We shall concentrate on the small amplitude regime and show that the dynamics is governed by the KP equation. It is also shown that two-soliton resonances could be obtained if two dimensional effects are taken into account.

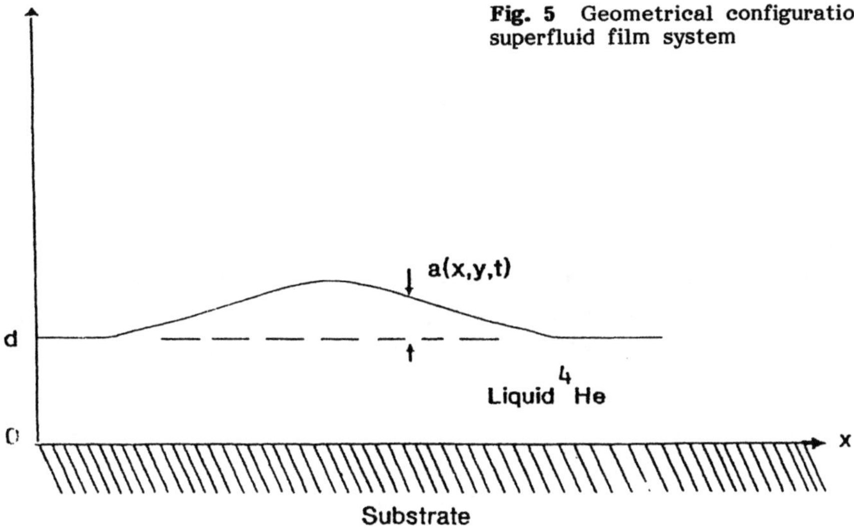

Fig. 5 Geometrical configuration of the thick superfluid film system

Since the superflow is irrotational we can describe it in terms of the velocity potential $\Phi(x,y,z,t)$. If the liquid is incompressible, Φ satisfies the Laplace's equation

$$\nabla^2\Phi = 0 . \tag{18}$$

The condition that the liquid does not penetrate into the substrate is expressed by the kinematical condition

$$\left.\frac{\partial\Phi}{\partial z}\right|_{z=0} = 0 . \tag{19}$$

On the film vapour interface we have

$$\frac{\partial z_1}{\partial t} + \left(\frac{\partial\Phi}{\partial x}\right)_1 \frac{\partial z_1}{\partial x} + \left(\frac{\partial\Phi}{\partial y}\right)_1 \frac{\partial z_1}{\partial y} - \left(\frac{\partial\Phi}{\partial z}\right)_1 = 0 . \tag{20}$$

The index 1 refers to the quantities on the film vapour interface, $z_1 = d + a(x,y,t)$ where $a(x,y,t)$ is the departure of the film surface from its equilibrium position. The equation of motion on the surface is [12]

$$\left(\frac{\partial\Phi}{\partial t}\right)_1 + \frac{1}{2}(\nabla\varphi)^2 - \frac{\sigma}{\rho}\left(\frac{\partial^2 a}{\partial x^2} + \frac{\partial^2 a}{\partial y^2}\right) + g_1 a - \frac{1}{2}\frac{g_2}{d}a^2 = 0 . \tag{21}$$

The last two terms in Eq. (21) represent the leading terms in the expansion of the van der Waals force term, $g_1 = 3\alpha/d^4$ and $g_2 = 12\alpha/d^4$, α being the van der Waals constant. $\bar{\rho}$ is the density and σ the surface tension. We look for a solution of the form [13] appropriate to a film

$$\Phi(x,y,z,t) = \sum_{n=0}^{\infty} z^n \varphi_n(x,y,t) . \tag{22}$$

From (18) and (19) we can formally write

$$\Phi(x,y,z,t) = \cos(z\nabla)\varphi_0(x,y,t) , \tag{23}$$

where ∇ is the gradient operator in two dimension. Using Eqs. (20) and (21) we study the dynamics of localized disturbances of long wavelength and of small amplitude in the superfluid film thickness. We choose the coordinate system in such a way that the x-axis is along the principal direction of propagation. We make the following transformation [5]:

$$X = x + C_3 t ; \quad t \rightarrow t . \tag{24}$$

We use the perturbation method of Tanuiti and Wei [14] wherein the following scale transformation is used.

$$\bar{x} = \varepsilon^{1/2} X ; \quad \bar{y} = \varepsilon y ; \quad \bar{t} = \varepsilon^{3/2} t \tag{25}$$

where ε is an infinitesimal. We now expand a and φ_0 in powers of ε.

$$a = a_0 + \varepsilon a_1(x,y,t) + \varepsilon^2 a_2(x,y,t) + \ldots \tag{26}$$

$$\varphi_0 = \varepsilon^{1/2}\varphi_0^{(1)}(x,y,t) + \varepsilon^{3/2}\varphi_0^{(2)}(x,y,t) + \ldots . \tag{27}$$

Substituting these in Eqs. (20) and (21) and comparing the coefficients of $\varepsilon^{3/2}$ and $\varepsilon^{5/2}$ we get

$$C_3 \frac{\partial a_1}{\partial \bar{x}} + d \frac{\partial^2 \varphi_0^{(1)}}{\partial \bar{x}^2} = 0 , \tag{28}$$

$$C_3 \frac{\partial^2 \varphi_0^{(1)}}{\partial \bar{x}^2} + g_1 \frac{\partial a_1}{\partial \bar{x}} = 0 , \tag{29}$$

$$C_3 \frac{\partial^2 \varphi_0^{(2)}}{\partial \bar{x}^2} - \frac{1}{2} C_3 d^2 \frac{\partial^2 \varphi_0^{(1)}}{\partial \bar{x}^4} + \frac{\partial}{\partial \bar{t}}\left(\frac{\partial \varphi_0^{(1)}}{\partial \bar{x}}\right) + \frac{\partial \varphi_0^{(1)}}{\partial \bar{x}} \frac{\partial^2 \varphi_0^{(1)}}{\partial \bar{x}^2}$$

$$- \frac{\sigma}{\rho} - \frac{\partial^3 a_1}{\partial \bar{x}^3} + g_1 \frac{\partial a_2}{\partial \bar{x}} - \frac{g_2}{d} a_1 \frac{\partial a_1}{\partial \bar{x}} = 0 , \tag{30}$$

$$\frac{\partial a_1}{\partial \bar{t}} + C_3 \frac{\partial a_2}{\partial \bar{x}} + \frac{\partial \varphi_0^{(1)}}{\partial \bar{x}} \frac{\partial a_1}{\partial \bar{x}} + a_1 \frac{\partial^2 \varphi_0^{(1)}}{\partial \bar{x}^2} + d \frac{\partial^2 \varphi_0^{(2)}}{\partial \bar{x}^2}$$

$$- d \frac{\partial^2 \varphi_0^{(1)}}{\partial \bar{y}^2} - \frac{d^3}{6} \frac{\partial^4 \varphi_0^{(1)}}{\partial \bar{x}^4} = 0 . \tag{31}$$

Using the boundary conditions we get

$$\frac{\partial \varphi_0^{(1)}}{\partial \bar{x}} = - \frac{C_3}{d} a_1 = - \frac{g_1}{C_3} a_1 ,$$

therefore $C_3^2 = g_1 d$. (32)

Eliminating $\frac{\partial \varphi_0^{(2)}}{\partial \bar{x}}$ between (30) and (31) we get

$$\frac{\partial}{\partial \bar{x}} \left[2C_3 \frac{\partial a_1}{\partial \bar{t}} + \left(g_2 - \frac{3C_3^2}{d} \right) a_1 \frac{\partial a_1}{\partial \bar{x}} + \left(\frac{\sigma d}{\bar{\rho}} - \frac{C_3^2 d^2}{3} \right) \frac{\partial^3 a_1}{\partial \bar{x}^3} \right]$$

$$- C_3^2 \frac{\partial^2 a_1}{\partial \bar{y}^2} = 0 . \tag{33}$$

Equation (33) can be reduced to the standard KP equation by setting

$$a_1 = -6\gamma\rho ; \quad \frac{1}{\gamma} = \left(g_2 - \frac{3C_3^2}{d} \right) \frac{1}{2C_3} ; \quad \xi = -k_0 \bar{x}$$

$$- \frac{1}{k_0^2} = \left(\frac{\sigma d}{\bar{\rho}} - \frac{C_3^2 d^2}{3} \right) \frac{1}{2C_3} ; \quad \eta = \sqrt{2k_0^2/C_3} \; \bar{y} , \; \tau = k_0 \bar{t} . \tag{34}$$

Using Eq. (34), Eq. (33) becomes

$$\frac{\partial}{\partial \xi} \left[\frac{\partial \rho}{\partial \tau} + 6\rho \frac{\partial \rho}{\partial \xi} + \frac{\partial^3 \rho}{\partial \xi^3} \right] + \frac{\partial^2 \rho}{\partial \eta^2} = 0 \tag{35}$$

which is the standard form of the KP equation with negative dispersion. This has been discussed in detail by Ohkuma and Wadati [15]. This is known to possess N soliton solutions [16]. The one soliton solution can be written as [17]

$$\rho = \frac{1}{2} k^2 \operatorname{sech}^2 \zeta , \tag{36}$$

$$\zeta = \frac{1}{2} k \left[\xi + p\eta - (k^2 + p^2)\tau \right] + \zeta^{(0)} . \tag{37}$$

Two Soliton Resonances

The two soliton solution of Eq. (35) is given by

$$\rho = 2(\log f_2)_{\xi\xi} \tag{38}$$

where

$$f_2 = 1 + \exp(2\zeta_1) + \exp(2\zeta_2) + A_{12}\exp[2(\zeta_1+\zeta_2)] , \tag{39}$$

$$\zeta_i = \frac{1}{2} k_i [\xi + p_i\eta - \tau(k_i^2 + p_i^2)] + \zeta_i^{(0)} , \tag{40}$$

and

$$A_{12} = \frac{3(k_1 - k_2)^2 - (p_1 - p_2)^2}{3(k_1 + k_2)^2 - (p_1 - p_2)^2} . \tag{41}$$

Soliton resonances occur when $A_{12} = 0$ or ∞ [18], i.e., for $3(k_1 \pm k_2)^2 - (p_1 - p_2)^2 = 0$. The plus sign refers to the case $A_{12} = \infty$ and is called the plus resonance and the minus refers to $A_{12} = 0$ and is called the minus resonance. The resonant soliton can be written as [6]

$$\rho^{(1\pm 2)} = \frac{1}{2}(k_1 \pm k_2)^2 \operatorname{sech}^2(\zeta_1 + \zeta_2) . \tag{42}$$

If the resonances are to be observed in actual situations then the resonance condition given by $A_{12} = 0$ or ∞ should be consistent with the conservation laws. Tajiri and Nishitani [19] have shown that this is indeed true for the KP equation (35).

Conclusion

We have studied the density fluctuations in thin ^{4}He films and have numerically analysed the time evolution of an arbitrary initial profile. We have also shown that for saturated films, the thickness fluctuations obey KP equation and would show the phenomenon of two soliton resonance.

Acknowledgements

We wish to thank Prof. J. C. Parikh and Dr B. R. Sitaram of PRL for help during the computation. One of us (JS) would like to thank CSIR, New Delhi for financial assistance through fellowship.

REFERENCES

[1] B. A. Huberman, Phy. Rev. Lett. **41** (1978) 1389.
[2] S. Nakajima, S. Kurihara and K. Tohdoh, J. Low Temp. Phy. **39** (1980) 465.
[3] A. C. Biswas and C. S. Warke, Phy. Rev. **B22** (1980) 2581.
[4] J. E. Rutledge, W. L. Mcmillan, J. M. Michell and T. E. Washburn, Phy. Rev. **B18** (1978) 2155.
[5] A. C. Biswas and C. S. Warke, Phy. Rev. **B28** (1983) 6539.
[6] J. Sreekumar and V. M. Nandakumaran, Phy. Lett. **112A** (1985) 168.

[7] S. Kurihara, J. Phy. Soc. Jpn. **50** (1981) 3262.
[8] J. Sreekumar and V. M. Nandakumaran, Pramana (in press).
[9] Radha Balakrishnan, R. Sridhar, and R. Vasudevan, Phys. Rev. **B39** (1989) 174.
[10] S. Kurihara, J. Phys. Soc. Jpn. **11** (1981) 3801.
[11] S. Nakajima, K. Dohdoh and S. Kurihara, J. Low Temp. Phys. **41** (1980) 521.
[12] J. Sreekumar and V. M. Nandakumaran, Mod. Phys. Lett. B (in press).
[13] G. L. Lamb, Jr, **Elements of Soliton Theory** (Wiley, New York,1980).
[14] T. Tanuiti and C. C. Wei, J. Phy. Soc. Jpn. **73** (1968) 941.
[15] K. Ohkuma and M. Wadati, J. Phy. Soc. Jpn. **52** (1983) 749.
[16] J. Satsuma, J. Phy. Soc. Jpn. **40** (1976) 286.
[17] J. Satsuma and M. J. Ablowitz, J. Math. Phy. **20** (1979) 1496.
[18] Hirota and Ito, J. Phy. Soc. Jpn. **52** (1983) 744.
[19] M. Tajiri and T. Nishitani, J. Phy. Soc. Jpn. **51** (1982) 250.

Studies on a Josephson Junction with Nonlinear Resistance

G. Ambika and K. Babu Joseph

Department of Physics, Cochin University of Science and Technology, Cochin 682022, India

We consider the model equation for the Josephson junction in the resistively shunted model (RSJ) with a periodic driving force and a resistance that varies inversely as voltage. This equation provides better agreement with experimentally observed I-V curves especially in the context of junctions using high T_c superconductors. The presence of the quadratic damping in our model is found to destabilise the system and introduce extremely long-lived transients. The nature of this metastable state is investigated and its departure from the usual RSJ model is highlighted. Chaos is found to set in due to the random shuttling between rotational and oscillatory modes in the system whose basins of attraction overlap. Stable limit cycles occur in a very narrow frequency range corresponding to large values of the damping factor.

1. Introduction

The Josephson junction is a familiar condensed matter system, capable of exhibiting deterministic chaos. The equation of motion representing this system is highly nonlinear with a damping that can be controlled easily and therefore serves as a model system in the study of chaotic behaviour in nonlinear dissipative systems. Thus the resistively shunted junction (RSJ) model may be used in many interesting physical situations like a damped driven pendulum [1], a particle moving viscously in a driven periodic potential [2], transport in charge density wave systems [3], etc. Since such a junction can be fabricated using integrated circuit techniques, a direct experimental study of chaotic behaviour in these systems is possible apart from numerical studies using a computer. Such investigations are important from the practical point of view also, since they give us an *apriori* idea about noise free regions which is useful while designing Josephson amplifiers. The study of chaos in this system began around the beginning of 1980, through the works of Huberman, Crutchfield [4-7], Kautz [8] and several others [9-12].

Research Reports in Physics Symmetries and Singularity Structures
Editors: M. Lakshmanan · M. Daniel

We present here numerical studies carried out on a modified Josephson junction. In the usual RSJ, the resistance is taken to be a constant. But we consider a resistance which depends inversely on voltage. This model is more realistic because it agrees better with the nonlinear nature of the I-V characteristics especially for junctions fabricated with high T_c superconductors [13]. The resulting equation of motion has a quadratic damping term, which is found to alter significantly the response of the system to the external periodic force. We try to analyse the nature of the chaotic behaviour exhibited by this system in various frequency ranges.

2. Driven Josephson Junction with Nonlinear Resistance

A simple model for the Josephson junction consists of the Josephson element shunted by a capacitance C and a resistance R (RSJ) [14]. When driven by a constant current I_{dc} and a periodic current of frequency ω, the equation for the circuit is,

$$C\frac{dV}{dt} + \frac{V}{R} + I_0 \sin\varphi = I_{dc} + I_1 \sin\omega t , \tag{1}$$

where $\varphi(t)$ is the junction phase and the junction voltage V is given by the Josephson relation,

$$\dot{\varphi} = 2eV/\hbar . \tag{2}$$

Using (2), equation (1) can be simplified to,

$$\ddot{\varphi} + k\dot{\varphi} + \sin\varphi = \rho_0 + \rho_1 \sin\omega t , \tag{3}$$

where the parameters k, ρ_0, and ρ_1 have been normalised in the conventional way with

$$k = (1/R)(\hbar/2eCI_0)^{1/2} ; \quad t \to t(\hbar C/2eI_0)^{1/2} ,$$

$$\rho_0 = I_{dc}/CI_0 \text{ and } \rho_1 = I_1/CI_0 . \tag{4}$$

Equation (3) has been the subject of extensive numerical and experimental studies during the past few years. The nature of the chaotic states as well as the possible mechanisms of their onset are now understood clearly [15,16].

In our model, we take the resistance R = constant/V and so (3) is modified as

$$\ddot{\varphi} + k|\dot{\varphi}|\dot{\varphi} + \sin\varphi = \rho_0 + \rho_1 \sin\omega t . \tag{5}$$

This equation can be solved analytically in the absence of the external driving and is known to possess periodic solutions [17]. With $\rho_1 = 0$ and $y = \dot{\varphi}$, (5) can be written as,

$$\frac{d(y^2)}{d\varphi} + 2k\, y^2 = 2\,\rho_0 - 2\,\sin\varphi\ , \tag{6}$$

which can be solved to give

$$y^2 = \rho_0/k + 4/(1 + 4k^2)^{1/2}\,\cos(\varphi+\beta) + C_1\,\exp(-2k\varphi) \tag{7}$$

with $\tan\beta = 2k$. The steady state solutions after the transients have died out is

$$y^2 = y_0^{1/2}(\rho_0 + \rho_2\,\cos(\varphi+\beta)) \tag{8}$$

with $\rho_2 = 2k/(1 + 4k^2)^{1/2}$ and $y_0 = k^{-1/2}$. The expressions for $\dot{\varphi}$ and φ can be obtained in terms of Jacobian elliptic functions.

But when driven by an external periodic force, the system has a tendency to resort to chaotic behaviour. The method of Melnikov analysis has been used to predict the lowest threshold for chaos in this system [18]. However, no detailed investigations regarding the possible routes to chaos, the nature of the chaotic state, etc., are known to date.

3. Low Frequency Response of the System

For convenience in our analysis, we rewrite (5) as a system of three first order equations. With $\varphi = x_1$, we have

$$\dot{x}_1 = x_2$$

$$\dot{x}_2 = -k|x_2|x_2 - \sin x_1 + \rho_0 + \rho_1 \sin x_3$$

$$\dot{x}_3 = \omega\ . \tag{9}$$

This constitutes a three-dimensional dissipative system and the rate at which the volume contracts is given by

$$\Lambda = -\,2k|x_2|\ . \tag{10}$$

This is interesting because in most of the nonlinear systems studied in the literature, Λ depends only on the parameters of the problem, but here it depends on x_2 as well. So richer dynamical behaviour is to be expected in this system. We choose those parameter values for which this richness of behaviour is explicit.

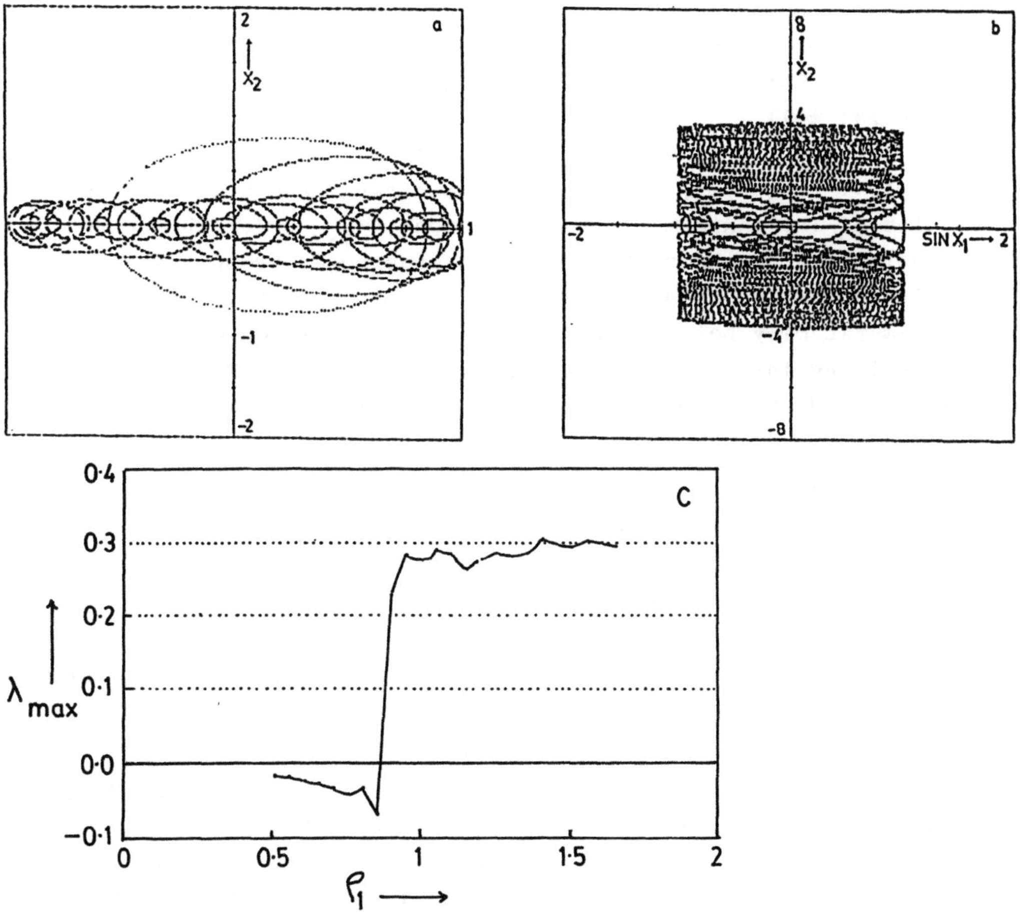

Fig. 1 Screw type motion existing in the low frequency region is plotted in (a) in the x_2-sin x_1 plane. The parameters are $\omega = 0.05$, $k = 0.1$, $\rho_0 = 0$ and $\rho_1 = 0.9$; (b) shows the chaotic trajectory of the system when ρ_1 is increased to 1.2. The variation of the maximum Lyapunov exponent λ_{max} with driving amplitude ρ_1 at $\omega = 0.07$ is given in (c).

The response of the system is found to depend primarily on the frequency of the applied force. At very low frequencies, i.e., $0.02 < \omega < 0.1$, we find that the system responds adiabatically to the driving force and experiences motions that have structure on a very fine scale. The trajectory in the x_1-x_2 plane follows a screw type behaviour around the x_1 axis, which is being repeated with the period of the applied force over a certain region. The amplitude of this motion decreases slowly but the system refuses to settle down to a stable trajectory. However, the phase amplitude φ is confined to the central region or potential well corresponding to $-\pi/2 < \varphi < \pi/2$. The pitch of the screw depends on the amplitude of the driving force ρ_1. For a given ω, the pitch increases with ρ_1 until for a critical value ρ_c the trajectory escapes from the central well.

The nature of the motion in two different time scales is evident from the typical trajectory in this parameter range shown in Fig. 1a. A computation of the maximum Lyapunov exponent, λ_{max} using the conventional technique [19], shows that λ_{max} remains negative but very small until the trajectory escapes from the well. After escape, the trajectory switches randomly between the rotational modes through damped oscillations, resulting in a chaotic orbit given in Fig. 1b. The variation of λ_{max} with the driving amplitude ρ_1 is shown in Fig. 1c.

4. Metastable Chaos

It is found that the screw type structure of the motion continues upto $\omega < 0.2$. For $\omega > 0.2$ the motion is of the oscillatory type. But in this region, asymptotically stable orbits are not found to occur. What is observed is a state of metastable chaos or preturbulence characterised by a trajectory that continues indefinitely for an extremely long interval of time. The system is incapable of stabilising into a limit cycle and this leads to broadening of peaks in the power spectrum computed using the Fast Fourier Transform (FFT) [20]. The λ_{max} corresponding to this behaviour is very small, either positive or negative. If λ_{max} is negative, the trajectory ultimately will settle down to a limit cycle and if it is positive, the final state would be a chaotic attractor. The small value of λ_{max} is found to be due to the fluctuations between convergence and divergence in the metastable state.

For $\omega = 0.6$, the metastable trajectory at $\rho_1 = 0.3$ is given in Fig. 2a. The FFT for this region shows two prominent but broad peaks which can be attributed to the small fluctuations associated with the transient nature of the motion (Fig. 2b). The corresponding Poincaré map reveals a slowly moving curve. The metastable state is observed to be sensitive to small changes in external disturbances and the time step of integration [21]. After a very long time of 500T, where T is the period of the driving term, the trajectory in Fig. 2a settles down to a limit cycle. However, if ρ_1 is close to ρ_c, the final state is the chaotic attractor. For $\rho_1 > \rho_c$, the system escapes to the chaotic state of the type described earlier. The chaotic trajectory and the corresponding FFT are shown in Figs. 2c,d.

As ω is increased further, the trajectories with positive λ_{max} exist even before escape from the well. This would mean that the state of

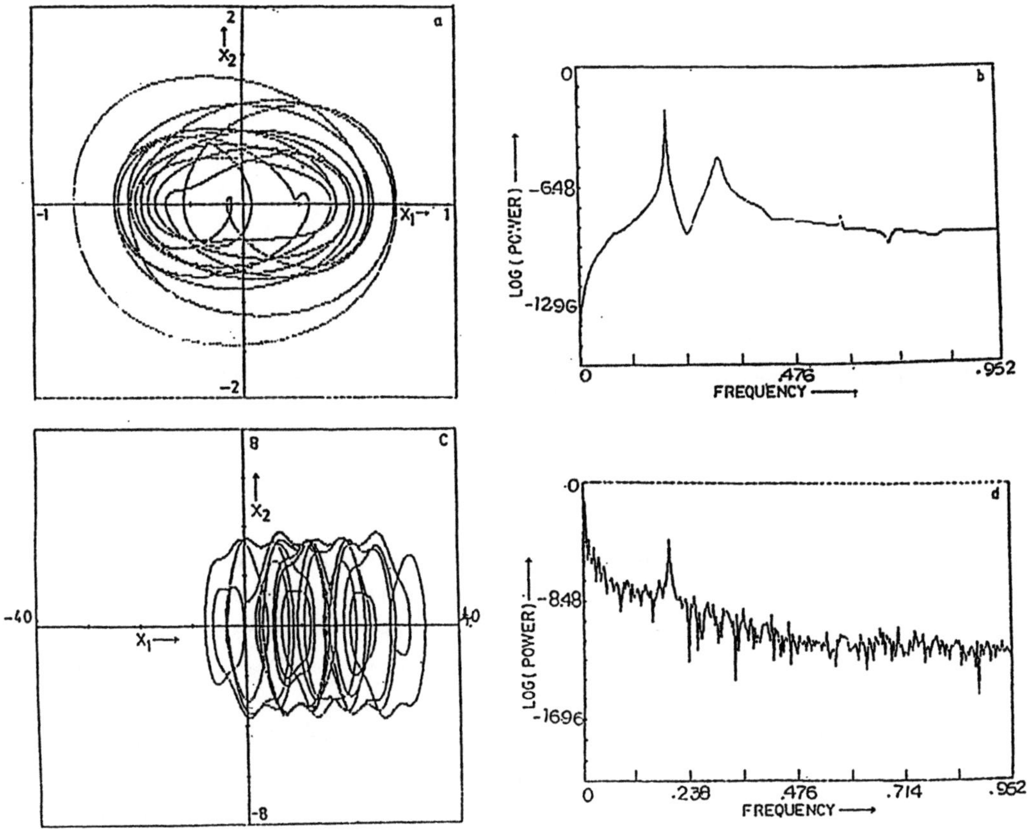

Fig. 2 The metastable trajectory in the x_1-x_2 plane for $\omega = 0.8$, $\rho_1 = 0.3$, $\rho_0 = 0$ and $k = 0.1$ is shown in (a) while (b) gives the corresponding power spectrum using FFT. The chaotic trajectory that sets in after escape to rotational modes with $\rho_1 = 2$ is seen in (c) and its FFT in (d).

metastable chaos would finally degenerate to a chaotic attractor. The variation of λ_{max} with ρ_1 in this frequency region is slow as is clear from Fig. 3.

Our attempts to isolate a sequence of period doubling bifurcations met with failure and the system is found to be incapable of entering into phase locked states also, except for a narrow region in frequency, $4 < \omega < 5.5$, corresponding to a winding number 3/2 and $k > 2$. With $\rho_0 \neq 0$, we find the behaviour of the system at all frequency ranges remains the same. But escape to rotational modes takes place at a much lower value of ρ_1.

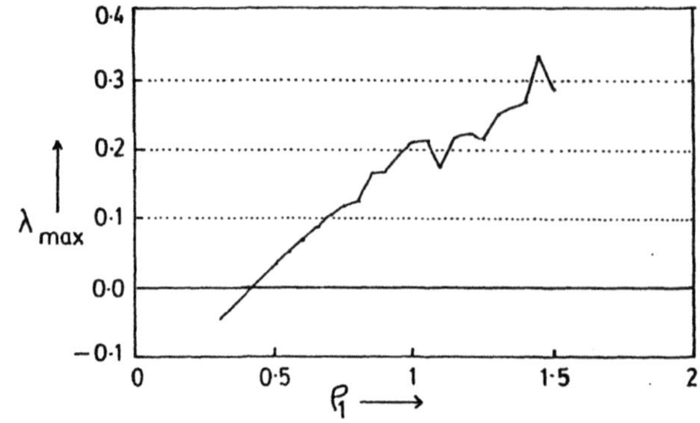

Fig. 3 The maximum Lyapunov exponent λ_{max} as a function of ρ_1 for ω = 1. We note that λ_{max} is positive before escape from the well.

5. Conclusion

We have investigated numerically the chaotic behaviour in a Josephson junction with quadratic damping. We find the mechanism of onset of chaos in this system is similar to that in the ordinary Josephson junction. Chaotic states arise mainly from the crisis where the basins of attraction of two attractors that have become unstable coalesce completely and covers the region of the attractor [16]. But in the usual RSJ, we have stable periodic limit cycles before crisis and they are found to undergo a period doubling route for ω = 0.6 [1]. In the system considered here, the rotational modes are unstable even before crisis and the oscillatory modes are not stable but metastable due to the continuous disturbance offered by the quadratic damping during each oscillation. Such metastable states have been reported recently in an actual junction [22], where their occurrence is attributed to the presence of noise in the system. But we feel this is mainly due to the nonlinear nature of the resistance. The role of noise is to shift the value of ρ_c at which the escape from the well takes place. In our model, the transition at ρ_c is from a metastable state to chaos where the basins of attraction of the two modes merge completely. The absence of a fully developed devil's staircase as well as the scarcity of phase locked states are other marked differences from the ordinary Josephson junction.

Acknowledgments

We are grateful to Dr V. M. Nandakumaran for informative discussions. One of us (GA) acknowledges the financial support of the Council of Scientific and Industrial Research (CSIR), New Delhi through a research associateship.

REFERENCES

[1] D. D. Humieres, M. R. Beasley, B. A. Huberman and A. Libchaber, Phys. Rev. **A26** (1982) 3483.
[2] W.-H. Steeb and A. Kunick, Phys. Rev. **A25** (1982) 2889.
[3] A. H. MacDonald and M. Plischke, Phys. Rev. **B27** (1983) 201.
[4] B. A. Huberman, J. P. Crutchfield and N. H. Packard, Appl. Phys. Lett. **37** (1980) 750.
[5] M. R. Beasley and B. A. Huberman, Comments Solid State Phys. **10** (1982) 155.
[6] B. A. Huberman and J. P. Crutchfield, Phys. Rev. Lett. **43** (1979) 1743.
[7] J. P. Crutchfield and B. A. Huberman, Phys. Lett. **A27** (1980) 407.
[8] R. L. Kautz and R. Monaco, J. Appl. Phys. **57** (1985) 875.
[9] M. Octavio, Phys. Rev. **B29** (1984) 1231.
[10] R. Kapral and P. Mandel, Phys. Rev. **A32** (1985) 1076.
[11] K. Wiesenfeld and N. F. Pedersen, Phys. Rev. **A36** (1987) 1440
[12] G. Cicogna, Phys. Lett. **A121** (1987) 403.
[13] S. P. Zhao, H. J, Tao, Y. F. Chen, G. C. Che and Q. S. Yang, Mod. Phys. Lett. **B3** (1989) 229.
[14] W. C. Stewart, Appl. Phys. Lett. **12** (1968) 277.
[15] E. G. Gwinn and R. M. Westervelt, Phys. Rev. Lett. **54** (1985) 1613.
[16] M. Iansiti, Q. Hu, R. M. Westervelt and M. Tinkham, Phys. Rev. Lett. **55** (1985) 746.
[17] N. F. Pedersen and K. Saermark, Physica **69** (1973) 572.
[18] M. Bartuccelli, P. L. Christiansen, N. F. Pedersen and M. P. Soerensen, Phys. Rev. **B33** (1986) 4686.
[19] W.-H. Steeb, J. A. Louw, **Chaos and Quantum Chaos** (World Scientific, Singapore, 1986), p. 42.
[20] E. O. Brigham, **The Fast Fourier Transform** (Prentice-Hall, New Jersey, 1974).
[21] Y. Morimoto, J. Phys. Soc. Japan **55** (1986) 2937.
[22] P. Silvestrini, O. Liengme and K. E. Gray, Phys. Rev. **B37** (1988) 1525.

Index of Contributors

Subject Index

B.N. Zakhariev, A.A. Suzko

Direct and Inverse Problems

Potentials in Quantum Scattering

1990. XIV, 223 pp. 42 figs. Softcover DM 48,– ISBN 3-540-52484-3

This textbook can almost be viewed as a "how-to" manual for solving quantum inverse problems, that is, for deriving the potential from spectra and/or scattering data. The formal exposition of inverse methods is paralleled by a discussion of the direct problem.
In part differential and finite-difference equations are presented side by side. A variety of solution methods is presented. Their common features and (dis)advantages are analyzed. To foster a better understanding, the physical meaning of the mathematical quantities are discussed in detail. Wave confinement in continuum bound states, resonance and collective tunneling, and the spectral and phase equivalence of various interactions are some of the physical problems covered.

J-M. Combes, A. Grossmann, P. Tchamitchian, Marseille (Eds.)

Wavelets

Time-Frequency Methods and Phase Space

Proceedings of the International Conference, Marseille, France, December 14-18, 1987

Inverse Problems and Theoretical Imaging

2nd rev. and enlarged ed. 1990. Approx. 350 pp. 98 figs. Softcover DM 98,– ISBN 3-540-53014-2

Time-frequency methods and phase space are well known to most physicists, engineers and mathematicians as is the traditional Fourier analysis. Recently the latter found for quite a few applications a competitor in the concept of wavelets. Crudely speaking a wavelet decomposition is an expansion of an arbitrary function into smooth localized contributions labeled by a scale and a position parameter.
This meeting brought together people exploring and applying these concepts in an interdisciplinary framework. The topics discussed range from purely mathematical aspects over signal analysis, seismic and acoustic applications via animal sonar systems to wavelets in computer vision.

P.C. Sabatier, Montpellier (Ed.)

Inverse Methods in Action

Proceedings of the Multicentennials Meeting on Inverse Problems, Montpellier, November 27th – December 1st, 1989

1990. XIV, 636 pp. 125 figs. Hardcover DM 138,– ISBN 3-540-51994-7

The basic idea of inverse methods is to extract from the evaluation of measured signals the details of the object emitting them. The applications range from physics and engineering to geology and medicine (tomography). Although most contributions are rather theoretical in nature, this volume is of practical value of interest to mathematicians. The review lectures and contributed papers are grouped into eight chapters dedicated to tomography, distributed parameter inverse problems, spectral and scattering inverse problems (exact theory), wave propagation and scattering (approximations); miscellaneous inverse problems and applications and inverse methods in nonlinear mathematics.

Research Reports in Physics

Manuscripts should be no less than 100 and no more than 400 pages in length. They are reproduced by a photographic process and must therefore be typed with extreme care. Corrections to the typescript should be made by pasting in the new text or painting out errors with white correction fluid. The typescript is reduced slightly in size during reproduction; the text on every page has to be kept within a frame of 16 × 25.4 cm ($6\frac{5}{16}$ × 10 inches). On request, the publisher will supply special stationary with the typing area outlined.

Editors or authors (of complete volumes) receive 5 complimentary copies and are free to use individual parts of the material in other publications later on.

All manuscripts, including proceedings, must contain a subject index. In the case of many-author books and proceedings an index of contributors is also required. Proceedings should also contain a list of participants, with complete addresses.

Our Instructions for the Preparation of Camera-Ready Manuscripts and further details are available on request.

Manuscripts (in English) or inquiries should be directed to

Dr. Ernst F. Hefter,
Physics Editorial 4,
Springer-Verlag, Tiergartenstrasse 17,
D-6900 Heidelberg, FRG,

(Tel. [0]6221-487495;
Telex 461723; Telefax 06221-43982).

Springer-Verlag
Berlin Heidelberg New York
London Paris Tokyo Hong Kong

GPSR Compliance

The European Union's (EU) General Product Safety Regulation (GPSR) is a set of rules that requires consumer products to be safe and our obligations to ensure this.

If you have any concerns about our products, you can contact us on ProductSafety@springernature.com

In case Publisher is established outside the EU, the EU authorized representative is:

Springer Nature Customer Service Center GmbH
Europaplatz 3
69115 Heidelberg, Germany

Batch number: 10584562

Printed by Printforce, the Netherlands